Mathematics for Finance, Business and Economics

Irénée Dondjio
Wouter Krasser

Eerste druk

Noordhoff Uitgevers Groningen/Houten

Ontwerp omslag: Rocket Industries, Groningen
Omslagillustratie: Getty Images

Eventuele op- en aanmerkingen over deze of andere uitgaven kunt u richten aan: Noordhoff Uitgevers bv, Afdeling Hoger Onderwijs, Antwoordnummer 13, 9700 VB Groningen, e-mail: info@noordhoff.nl

1 / 15

Deze uitgave is gedrukt op FSC-papier.

ISBN 978-90-01-81862-3
NUR 919

Preface

Welcome to your new Mathematics book. Our desire to write this book was borne out of our experience of lecturing Mathematics, Economics and Marketing courses for five years at The Hague University of applied Science. The motivation arose when a couple of our students approached us and suggested writing a simplified Mathematics syllabus because the notes we gave in class were more comprehensive and easier to understand than the textbook.

It occurred to us that some students with an intermediate or advanced level of Mathematics were strongly motivated to tackle the course, whilst some with no background or a very poor knowledge of Mathematics tended to be daunted by the complex text, and this demotivated them. In order to improve students' knowledge of Mathematics, we decided to write a book that would be simple to understand, and make the subject more appealing and less off-putting to them.

This book is written informally, in order to be easy to read and understand. It's not just a lot of text, and then lots of exercises, such as you might find in traditional textbooks but a combination of explanations, explorations, and real-life applications of major concepts.

Acknowledgments

This book would not exist were it not for the efforts of The Hague University; we would like to express our appreciation for their unconditional support. In particular we would like to thank Ms. Akebe and Mr. Mugabi for sharing their expertise on the Mathematics of Finance, and Mr. Schumacher for making this book possible in the first place. Our gratitude also goes out to our colleagues from Saxion, Fontys, and the Hogeschool van Amsterdam for providing help and support in innumerable ways. Furthermore we would like to thank Donna Scott for editing the manuscript. And last, but not least we would like to acknowledge and extend our heartfelt gratitude to our family and friends for their encouragement and patient love which have enabled us to complete this handbook. We would especially like to express our gratitude to Etienne Nankeu.

Autumn 2012, The Hague
Irénée Dondjio, MBA.
Wouter Krasser, MSc.

Table of Contents

Introduction

For a long time, Mathematics has been seen as subject for highly skilled and specialized students. Business students are often not able to understand the benefits of Mathematics beyond simple everyday calculations.

Mathematics develops the imagination. It requires clear and logical thought. Mathematics is a science of patterns and structures that provides students with a uniquely powerful set of tools for understanding and solving real-life problems. Math studies non-tangible objects like numbers, circles and functions. However, we can profit greatly from this science because it has a big impact on our daily lives. For example, Math is essential in medicine, for analysing data on the causes of illness and the use of new drugs. Travel by aeroplane would not be possible without the Mathematics of airflow and control systems. Furthermore, Math is very useful in Finance, Economics and Accounting. For example, in exercise 6.19: suppose you have €1000 in a savings account with a yearly interest rate of 3%; how long would you have to wait until your money has doubled? Or exercise 5.10: can we build a model of the behaviour of customers in order to predict the market for the next few years? These and many other questions will be answered in this book.

All Business students need to master the basic concepts of Mathematics, as this is the key to understanding other courses, such as Economics, Finance, Statistics, and Accounting. Therefore, this book follows a logical order (from easy topics to challenging topics) so that students can stay focused.
Because this book is mainly dedicated to Business and Economics students, we have tried to relate it to subjects such as Economics, Marketing and Finance through real-life problems and situations.

In this book you will find theory, examples and exercises in each chapter. Solutions to the exercises can be found at the back of the book. The corresponding computations and more exercise material can be found on the website www.mathforfinance.noordhoff.nl.

The truncated icosahedron, aka the football

Source: Jan van de Craats, Faculteit FNWI UvA

Mc. 14,45

1 Elementary Mathematical Concepts and Operations

After studying this chapter you should be able to:
- add, subtract, multiply and divide positive and negative numbers
- understand the concept of a square root
- expand and evaluate an algebraic expression
- plot points on a graph

The aim of this chapter is to introduce the basic operations (addition, substraction, multi plication and division) used in algebra. The main focus will be on the properties of real numbers and the basic manipulation of algebraic expressions, collecting algebraic expressions with like terms.

Arithmetic is the ABC of Math. Addition, subtraction, multiplication, and division are the basics of Math and every Math operation known to humankind. In one way or another, every equation, graph, and many other things can be broken down into the ABCs of Math: the four basic operations. As people say, Math is a language, and addition, subtraction, multiplication, and division are its alphabet, along with the number line as well. The properties are basically proven ways to apply the mechanics of arithmetic to certain situations (source: www.writework.com).

1

Is Mathematics only for a few kids?

Is Mathematics only for a few clever kids who are genetically disposed towards the subject? Many teachers believe, either explicitly or implicitly, that some children may be born with mathematical aptitudes or mathematics genes, and others are not. Some teachers even believe that children from certain groups (bases on factors such as gender, ethnicity and race) are blessed with superior mathematical ability. Some teachers feel there is not much that can be done to change or improve the innate ability of those unfortunate children who are inherently not good at mathematics.

The mathematical interests and knowledge young children bring to school may indeed differ, but the causes are more likely to be their varying experiences, rather than their biological endowment. While teachers should be aware of and sensitive to these differences, they should never lose sight of the fact that all children, regardless of their backgrounds and prior experiences, have the potential to learn Mathematics. In fact, the gaps in early Mathematics knowledge can be narrowed or even closed by good Mathematics curricula and teaching. Teachers should strive to hold high expectations and support for all children, without any ungrounded biases. When a teacher expects a child to succeed (or fail), the child tends to live up to that expectation.

Source: www.earlychildhoodaustralia.org

1.1 Simple Operations on Positive and Negative Numbers

FIGURE 1.1 An Arabic public telephone keypad

The digits 1, 2, etc. originate from Arabic numerals. Before adopting these symbols, Europeans used Roman numerals I, II, III, IV, V, etc. Negative numbers were not fully accepted until around 1800. In this book, we will use the decimal point and the thousand separator in our figures, like this:

$$12{,}345.67$$

The above means twelve thousand, three hundred forty-five and sixty-seven hundredths.

FIGURE 1.2 The number line illustrating the order of numbers

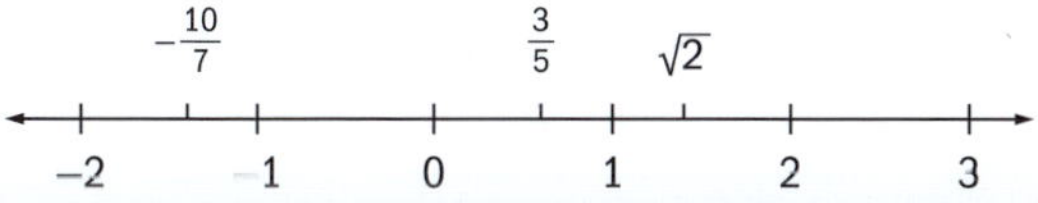

1

In Math there are positive and negative numbers, and the number 0, which is neither positive nor negative. We can think of a positive number as a credit and a negative number as a debt. The order of numbers is illustrated in figure 1.2, the picture of the number line, for example: $2 > 1$, $-2 < -1$, and $1 > -2$. Numbers can be added, subtracted and multiplied. All numbers can also be divided, but not by zero.
An example of the addition of positive and negative numbers goes like this: suppose you have a debt of €300. That means the balance of your bank account equals –€300. If you deposit €400 in your bank account, your new balance is $-300 + 400 = 100$ euros. Or suppose again you have a debt of €300, and your boyfriend has a debt of €200. So the balances of both of your bank accounts are –€300 and –€200 respectively. Now after your wedding you decide to join both of your bank accounts. Together you have a debt of €500, and the balance of your joint account equals $-300 + -200 = -500$ euros.

The multiplication and division of positive and negative numbers are illustrated in Table 1.1.

TABLE 1.1 Multiplication and division of positive and negative numbers

× or ÷	+	–
+	+	–
–	–	+

Example 1.1

Multiplication and Division of Positive and Negative Numbers Referring to Table 1.1 we see that:
$(+1)\times(+2) = 2$,
$(+1)\times(-2) = -2$,
$(-1)\times(+2) = -2$, and
$(-1)\times(-2) = 2$.

The same holds for division, i.e.

$\frac{+1}{+2} = +\frac{1}{2}$,

$\frac{+1}{-2} = -\frac{1}{2}$,

$\frac{-1}{+2} = -\frac{1}{2}$, and

$\frac{-1}{-2} = \frac{1}{2}$.

N.B. When there is no sign in front of the number, it is always positive. I.e. $3 = +3$.

EXTEND YOUR KNOWLEDGE
Do you know why $(-1)\times(-1)=1$?
The axioms of arithmetic state that for all numbers x, y, z we have:
a $x+-x=0$,
b $x\times 0=0$,
c $x\times(y+z)=x\times y+x\times z$ (distributivity), and
d $(-1)\times x=-x$.

Now:

$$\begin{aligned} 1+-1&=0 \text{ (by axiom a)},\\ (-1)\times(1+-1)&=(-1)\times 0 \text{ (by axiom b)},\\ (-1)\times 1+(-1)\times(-1)&=0 \text{ (by axiom c)},\\ -1+(-1)\times(-1)&=0 \text{ (by axiom d)},\\ \text{Hence } (-1)\times(-1)&=1. \end{aligned}$$

1.2 Fractions and Square Roots

A fraction is the ratio of two whole numbers, i.e.

$$\frac{6}{2}$$

is the ratio of 6 and 2. In this fraction 6 is called the *numerator*, and 2 the *denominator*. The ratio of 6 and 2 equals 3:

$$\frac{6}{2}=3,$$

for 6 is three times as big as 2:

$$6=3\times 2.$$

Using another example, we can also determine the ratio of $\frac{1}{2}$ and $\frac{1}{4}$:

$$\frac{\frac{1}{2}}{\frac{1}{4}}=2,$$

because $\frac{1}{2}$ is twice as big as $\frac{1}{4}$:

$$\frac{1}{2}=2\times\frac{1}{4}.$$

Two times a quarter equals a half may be demonstrated by: half a euro is twice as much as a quarter of a euro (50 cents is twice as much as 25 cents).

For calculations with fractions the following rules hold:

1

THEOREM 1.1

Calculation Rules for Fractions

Let $a, b, c,$ and d be real numbers, with b and d unequal to zero. Then:

- The *addition* of two fractions with the *same* denominator is defined by:

$$\frac{a}{d}+\frac{c}{d}=\frac{a+c}{d} \tag{1.1}$$

- The *addition* of two fractions with *different* denominators is defined by:

$$\frac{a}{b}+\frac{c}{d}=\frac{ad+bc}{bd} \tag{1.2}$$

- The *multiplication* of two fractions is defined by:

$$\frac{a}{b}\times\frac{c}{d}=\frac{ac}{bd} \tag{1.3}$$

- The *division* of two fractions is defined by:

$$\frac{\frac{a}{b}}{\frac{c}{d}}=\frac{a}{b}\times\frac{d}{c}=\frac{ad}{bc} \tag{1.4}$$

Equation (1.4) shows that dividing by a fraction is equivalent to multiplying by its inverse (the inverse of a number $x \neq 0$ is $\frac{1}{x}$). Note that we have used the notational convention ab for $a \times b$.

Here is an example of the application of equation (1.1): Suppose you have $\frac{1}{5}$ of a euro, 20 cents. You add $\frac{2}{5}$ of a euro, 40 cents. The result is 60 cents, $\frac{3}{5}$ of a euro:

$$\frac{1}{5}+\frac{2}{5}=\frac{1+2}{5}=\frac{3}{5}$$

Here is an example of the application of equation (1.2): Suppose you have $\frac{1}{2}$ of a euro, 50 cents. You add $\frac{3}{5}$ of a euro, 60 cents. The result is 110 cents, 1.1 euros:

$$\frac{1}{2}+\frac{3}{5}$$

Since two fractions with a different denominator cannot be added immediately, we use equation (1.2):

$$\frac{1}{2}+\frac{3}{5}=\frac{1\times5}{2\times5}+\frac{3\times2}{2\times5}=\frac{1\times5+3\times2}{10}=\frac{11}{10}=1.1$$

Here is an example of the application of equation (1.3):

$$\frac{1}{3}\times\frac{2}{5}=\frac{1\times2}{3\times5}=\frac{2}{15}$$

Here is an example of the application of equation (1.4):

$$\frac{\frac{1}{2}}{\frac{3}{4}}=\frac{1}{2}\times\frac{4}{3}=\frac{1\times4}{2\times3}=\frac{4}{6}=\frac{2}{3}$$

We will now turn to the definition of the square and the root of a number. The *square of a number* is that of a number multiplied by itself, and the *square root* answers the question 'Which number was multiplied by itself in order to get this new number?'. A square root is useful for solving an equation like $x^2 = 64$ (see figure 1.3).

Definition 1.1 Square
Let x be any number. The square of x, x^2 is defined by

$$x^2 = x \times x$$

For example: $5^2 = 5 \times 5 = 25$ and $(-3)^2 = (-3) \times (-3) = 9$.

FIGURE 1.3 The square of 8 equals 64 because $8 \times 8 = 64$

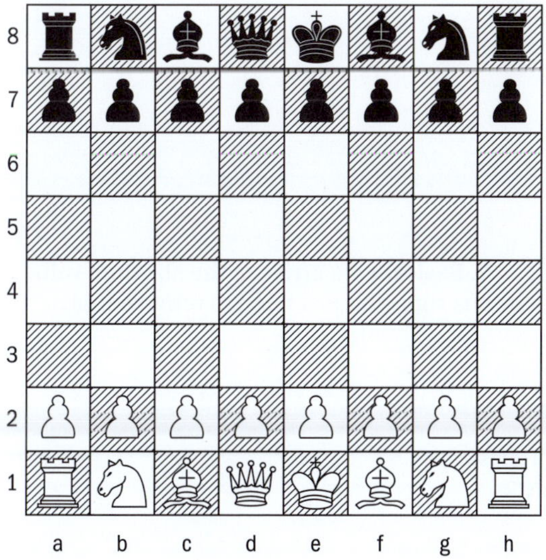

1

Definition 1.2 Square root
Let x be a non-negative number, then the square root of x,

$$y = \sqrt{x},$$

is that number $y \geq 0$ which, when squared, equals x, i.e.:

$$y^2 = (\sqrt{x})^2$$

Example 1.2

The square root of 9 equals 3, because $3^2 = 9$. Note also that $(-3)^2 = 9$, so you might think that -3 would also be the square root of 9, however, the definition of a square root states that only the non-negative value applies. Note that negative numbers do not have a square root. For example: $\sqrt{-9}$ does not exist because there is no number that, when squared, equals -9.

THEOREM 1.2

Calculation Rules for the Square Root

Let a, $b \geq 0$, $c > 0$, and d be any number. Then consider the following rules:

$$\sqrt{ab} = \sqrt{a}\sqrt{b} \quad (1.5)$$

$$\sqrt{\frac{a}{c}} = \frac{\sqrt{a}}{\sqrt{c}} \quad (1.6)$$

$$\sqrt{d^2} = |d| \quad (1.7)$$

$$(\sqrt{a})^2 = a \quad (1.8)$$

Where $|d|$ denotes the absolute value of d, as defined by:

$$|d| = \begin{cases} -d \text{ for } d < 0 \\ 0 \text{ for } d = 0 \\ d \text{ for } d > 0 \end{cases} \quad (1.9)$$

For example: $|-2| = 2$ and $|2| = 2$. Basically, to arrive at the absolute value of a number expressed with a minus sign, you just need to remove that minus sign.

Example 1.3

An example of equation (1.5) is:
$\sqrt{8} = \sqrt{4 \times 2} = \sqrt{4}\sqrt{2} = 2\sqrt{2}$

Example 1.4

An example of equation (1.7) and the absolute value (1.9) is:
$\sqrt{(-4)^2} = |-4| = 4$

1.3 BEDMAS

Just as it matters in which order you put on your shoes and socks, it matters in which order you add and multiply. The order in which mathematical operations are executed is often referred to as 'BEDMAS' (see figure 1.4).

FIGURE 1.4 Which of the calculators is right?

'BEDMAS' is an acronym that stands for:

- **B** brackets
- **E** exponents (means powers, squares and roots)
- **DM** divide and multiply, from left to right
- **AS** add and subtract, from left to right

Example 1.5

Let us calculate:

$$\frac{(3+6)-8\times 3}{24+6^2}$$

Following 'BEDMAS' we need to start with the **B**rackets:

$$(3+6)-8\times 3\div 24+6^2$$
$$=9-8\times 3\div 24+6^2$$

There is an **E**xponent, 6^2, which has to be done now:

$$9-8\times 3\div 24+6^2$$
$$=9-8\times 3\div 24+36$$

We continue with **D**ivision and **M**ultiplication from left to right:

$$\begin{aligned} & 9-8\times 3\div 24+36 \\ =\ & 9-24\div 24+36 \\ =\ & 9-1+36 \end{aligned}$$

and we finish with **A**ddition and **S**ubtraction from left to right:

$$\begin{aligned} & 9-1+36 \\ =\ & 8+36 \\ =\ & 44 \end{aligned}$$

1.4 Algebraic Expressions

Definition 1.3 Algebraic expression
An algebraic expression is made up of the signs, or numbers, and letters, or symbols of algebra. It is also composed of terms which can be variable or constant.

$$\overbrace{\underbrace{5}_{\text{coefficient}}\ \underbrace{x}_{\text{variable}}}^{\text{term}} - \overbrace{\underbrace{3}_{\text{constant}}}^{\text{term}}$$

Definition 1.4 Variable
Variables are unknown values that may change within the scope of a given problem or set of operations. Variables are represented by letters most of the time, and the ones often used are: x, y, and z. However any other letter could be used.

Definition 1.5 Constant
A constant is a fixed value that does not change, like 1, 2, 3 etc.

Example 1.6

Addition and Subtraction of Algebraic Expressions
How can we simplify an expression such as
$2x+4y-25x+30y-45xy+25yx$?

First we collect like terms; there are 3 different sorts, x, y, and xy:

x	y	xy
$2x-25x$	$4y+30y$	$-45xy+25yx$
$=-23x$	$=34y$	$=-20xy$
		Note that $xy=yx$.

After adding all the terms we get $-23x+34y-20xy$.

1.5 Expanding Brackets

Expanding brackets, otherwise known as evaluating an expression, involves removing the brackets in order to simplify the expression down to a single numerical value. We can expand the brackets of the following expression:

$$(x+3)(x-2)$$

FIGURE 1.5 The banana method

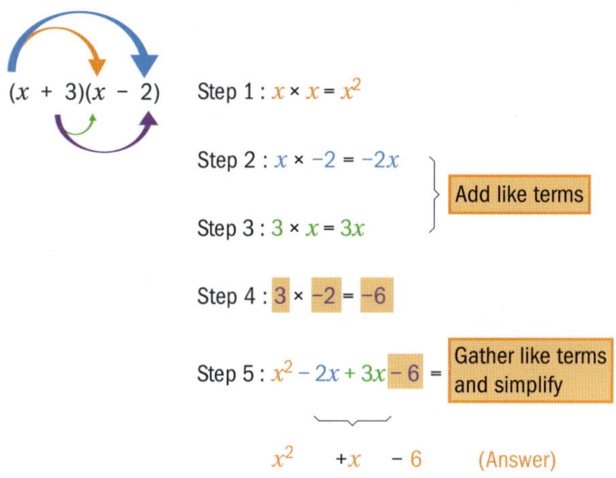

A convenient way of expanding these brackets is by using the *banana method*, see figure 1.5:

$$(x+3)(x-2)$$

$$\begin{aligned}(x+3)(x-2) &= x^2-2x+3x-6\\ &= x^2+x-6\end{aligned}$$

Does this make any sense? Let us check the following example. We already know that:

$$(1+2)(3+4)=(3)(7)=3\times 7=21 \tag{1.10}$$

Expanding the brackets first yields:

$$\begin{aligned}(1+2)(3+4) &= 3+4+6+8\\ &= 21\end{aligned}$$

Which is the same answer as in (1.10).

1.6 Factorizing Algebraic Expressions

Factorizing an algebraic expression is the opposite of expanding. You start with a sum or difference of terms and finish up with a product.

For example, by factorizing the following expression:

$$ab + ac$$

You get

$$a(b + c)$$

We have used the *common factor method* in the example: all terms have a as a factor, so a is the common factor. We should now use the brackets to write down the common factor outside the brackets as follows:

$$a(\quad)$$

To find out what goes inside the brackets, divide the original terms by the common factor(s).

For the example, divide the original terms by a:

$$\frac{ab}{a} = b$$

$$\frac{ac}{a} = c$$

The last step is to write these new expressions inside the brackets, like this:

$$a(b + c)$$

Example 1.7

Factorize $2yx + 4x$
$2yx + 4x$
$= 2xy + 2x \times 2$
$= 2x(y + 2)$

1.7 The Cartesian Plane

The Cartesian plane was named after the famous French philosopher and mathematician Rene Descartes (figure 1.6). When two perpendicular number lines intersect, a Cartesian plane is formed.

Rene Descartes, a French philosopher, viewed the world with a cold analytical logic. He saw all physical bodies, including the human body, as machines operated by mechanical principles. His philosophy was derived from the austere logic of 'cogito ergo sum' meaning: I think therefore I am.
In Mathematics, Descartes's chief contribution was in analytical geometry. Descartes's portrait is quadrisected by the axes of his great advance in analytical geometry: the Cartesian Plane. It enabled an algebraic representation of geometry (source: http://mathematicianspictures.com).

FIGURE 1.6 René Descartes (1596–1650)

FIGURE 1.7 The coordinate plane

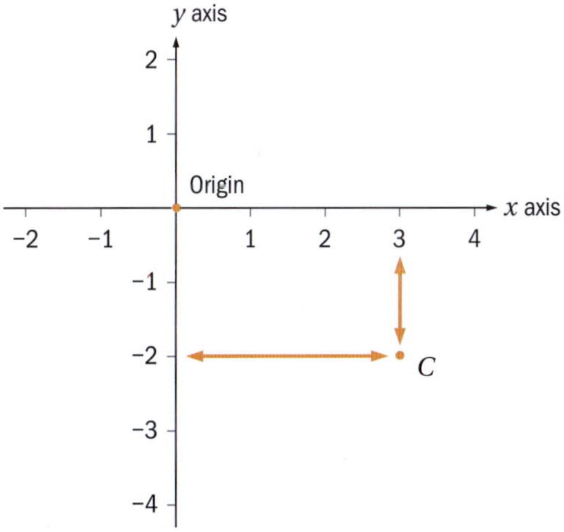

The Cartesian Plane is a plane with a rectangular coordinate system that associates each point in the plane with a pair of numbers. A coordinate plane has two axes: the x-axis (the horizontal line), and the y-axis (the vertical line). These two axes originate in the origin O with coordinates (0, 0). In figure 1.7 point C has coordinates (3, −2). 3 is the distance of this point from the origin in the direction of the x-axis, and −2 is the distance of this point from the origin in the direction of the y-axis, hence the line goes downwards because −2 is negative.

Summary

- Multiplication and division of positive and negative numbers are illustrated in this table:

× or ÷	+	−
+	+	−
−	−	+

- Rules for calculations with fractions are:
  - $\frac{a}{d}+\frac{c}{d}=\frac{a+c}{d}$
  - $\frac{a}{b}+\frac{c}{d}=\frac{ad}{bd}+\frac{bc}{bd}=\frac{ad+bc}{bd}$
  - $\frac{a}{d}\times\frac{c}{d}=\frac{ac}{bd}$
  - $\frac{\frac{a}{b}}{\frac{c}{d}}=\frac{a}{b}\times\frac{d}{c}=\frac{ad}{bc}$
- $x^2=x\times x$
- $y=\sqrt{x}$ shows that non negative number y satisfies $y^2=x$
- Rules for calculations with square roots are:
  - $\sqrt{ab}=\sqrt{a}\sqrt{b}$
  - $\sqrt{\frac{a}{c}}=\frac{\sqrt{a}}{\sqrt{c}}$
  - $(\sqrt{a})^2=a$
  - The square root of a negative number does not exist.
  - $\sqrt{d^2}=|d|$,

where $|d|$ denotes the absolute value of d, as defined by

$$|d|=\begin{cases}-d \text{ for } d<0\\ 0 \text{ for } d=0\\ d \text{ for } d>0\end{cases}$$

- BEDMAS tells you the order to follow: brackets, exponents, division and multiplication (from left to right), addition and subtraction (from left to right)
- Expanding brackets:

$$(a+b)(c+d)=ac+ad+bc+bd$$

- What is the difference between variable, constant and coefficient?

Exercises

1

Complete these exercises without using a calculator:

1.1 You borrowed €1,000 from the bank last month to buy your books, and this month you paid back €800. But you realize that you need a notebook, so you borrow another €500 to buy one. What is the total amount that you owe the bank now?

1.2 Last week the balance of your bank account was €600 in debit and this week you withdrew €300. How much is your account now in debit by?

1.3 Determine the validity of the following statements. Are they true or false?

a $-3 \leq 4$
b $8 < 3$
c $-8 < -3$
d $0.7 \geq -4$
e $-5.3 > -7$
f $\frac{3}{4} < \frac{6}{8}$
g $-\frac{3}{4} \leq -\frac{6}{8}$
h $1+2=2+1$
i $1-2=2-1$

1.4 Calculate

a $\frac{1}{3}+\frac{1}{3}$
b $\frac{1}{3}+\frac{2}{3}$
c $\frac{1}{3}+\frac{1}{6}$
d $\frac{1}{2}-\frac{1}{4}$
e $\frac{1}{2}+\frac{1}{3}$
f $\frac{1}{2}-\frac{2}{3}$
g $\frac{1}{3}+\frac{3}{2}$
h $\frac{5}{7}-\frac{7}{5}$

1.5 Calculate

a $\frac{1}{2}\times\frac{1}{2}$

b $\frac{1}{3}\times\frac{2}{1}$

c $\frac{1}{3}\times 2$

d $\frac{2}{3}\times\frac{-1}{2}$

e $\frac{8}{-7}\times\frac{3}{5}$

f $\frac{7}{8}\times\frac{-5}{3}$

g $\frac{3}{5}\times\frac{5}{2}$

h $\frac{5}{7}\times\frac{7}{5}$

1.6 Calculate

a $3+(6-2)$

b $5-(8-7)$

c $15-(4-2)$

d $(17-1)-4$

e $23-(4-2)$

f $5+(14-7)-3$

g $4+(71-1)+1$

h $16-(8-7)-5$

1.7 Calculate

a $\frac{5}{4}\times\frac{7}{8}$

b $\frac{18}{7}+\frac{3}{8}$

c $-\frac{5}{7}+\frac{7}{5}+\frac{3}{2}$

d $-\left(\frac{7}{5}+\frac{5}{9}\right)+\left(-\frac{3}{9}+\left(-\frac{14}{5}\right)\right)$

e $-\left(\frac{10}{5}-\frac{5}{8}\right)-\left(-\frac{3}{9}-\left(-\frac{12}{5}\right)\right)$

f $\frac{\frac{1}{2}}{3}$, hint: $3=\frac{3}{1}$

g $\frac{1}{\frac{2}{3}}$

h $\frac{\frac{1}{4}}{\frac{2}{3}}$

i $\frac{1}{\frac{1}{2}}$

j $\dfrac{\frac{1}{1}}{2}$

k $\dfrac{\frac{1}{2}}{\frac{1}{3}}$

l $\dfrac{\frac{1}{3}}{\frac{1}{2}}$

1.8 Calculate

a $(1-2)-3$
b $1-(2-3)$
c $2\times3+4$
d $2\times(3+4)$
e $10-5+8$
f $10-(5+8)$
g $10\div2+3$
h $10\div(2+3)$

1.9 Evaluate and simplify

a $\sqrt{9}$
b $\sqrt{27}$
c $\sqrt{4\times9}$
d $\sqrt{(-9)^2}$
e $\sqrt{\dfrac{9}{16}}$
f $\sqrt{\dfrac{27}{8}}$
g $(\sqrt{3})^2$
h $\sqrt{3^2}$

1.10 Calculate

a $(7+6)^2\div13+1$
b $(1+2)^3+4\div5-6\times7$
c $(3+5-7)\times6\div2+3^2$
d $(3+5)-7\times6\div2+3^2$
e $1\div\dfrac{2}{3}$
f $\dfrac{1}{2}\div3$
g $\dfrac{1}{2}+\dfrac{3}{4}$
h $\dfrac{1+2}{3+4}$
i $1+2\div3+4$
j $\dfrac{\frac{1}{2}}{3}+4$

1.11 Referring to figure 1.4, which calculator is right according to BEDMAS?

1.12 Calculate

a $(\sqrt{3}-\sqrt{2})(\sqrt{3}+\sqrt{2})$

b $2\sqrt{6}-\sqrt{2}\sqrt{3}$

c $\dfrac{\sqrt{4}}{2}$

d $\sqrt{\dfrac{4}{2}}$

e $\sqrt{4+5}$

f $\sqrt{4}+5$

g $\sqrt{3\times3+20-8\div2}$

h $\sqrt{3\times3}+\sqrt{20-8\div2}$

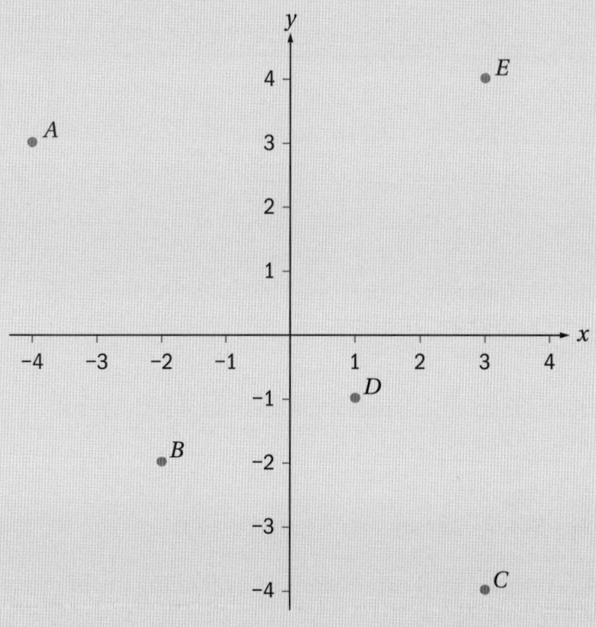

1.13 Use this coordinate plane to answer the following questions:

a Which point has coordinates $(3, -4)$?

b What are the coordinates of the other points $A, B, \ldots, E$?

c Is there a point plotted at coordinates $(1, -1)$?

1.14 Plot the following points in this coordinate plane.

$A=(4, 0)$ $B=(1, 2)$ $C=(2, 1)$ $D=(-3, 3)$
$E=(-4, 0)$ $F=(-1, -2)$ $G=(2, -1)$ $H=(0; 0.7)$
$I=(-0.8; 1)$ $J=\left(-3\frac{1}{2}, 1\right)$ $K=\left(\frac{5}{2}, \frac{7}{2}\right)$ $L=\left(-\frac{18}{17}, \frac{22}{7}\right)$

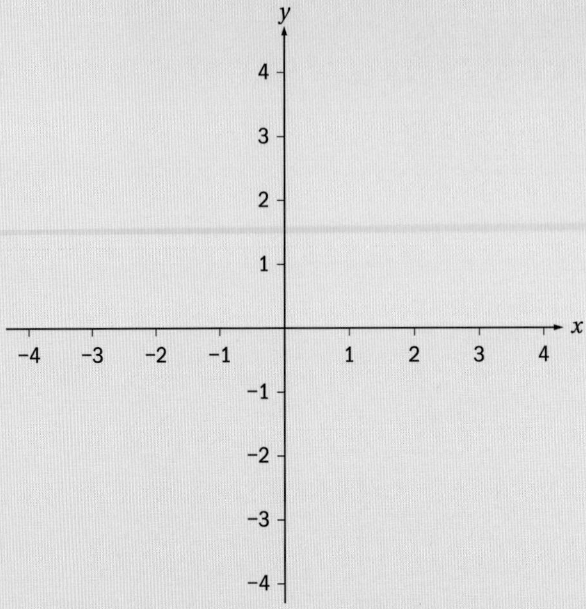

1.15 Simplify the following expressions by collecting like terms:

a $8x-4x+7y-5x+x+3y-10y$
b $5u-7v+6uv-5vu+8v-4u$

1.16 Expand the following expressions and simplify the result.

a $7(x+2y)+8(2x-y)$
b $10(p+q)-3(p-q)$
c $4(x+6)+5(2x+6)$
d $-4(-2x+5)+5(-3x+6)$

1.17 Simplify the following expressions by expanding the brackets and adding like terms: [see fig 1.4]

a $(x+7)(x-4)$
b $(x-y)(x+y)$
c $(\sqrt{a}+\sqrt{b})(\sqrt{a}-\sqrt{b})$
d $(x+y)^2$
e $(2b+6)(b-7)$
f $(2x+5y)+(3x-2y)$
g $(7.5x^2-x-4)+\left(\frac{1}{2}x^2-2x-3\right)+\frac{4}{2}x^2$

1.18 Show that
$\sqrt{2}+\sqrt{3}=\sqrt{5+2\sqrt{6}}$

1.19 Factorize the following expressions:

a $3x+3$
b $(x+1)(x+5)+7(x+1)$
c $(x+1)(3x+4)+(x+1)(x-3)$
d $(x+1)(4x+9)-5(x+1)$

2
Linear Equations

After studying this chapter you should be able to:

- solve linear equations and graph them
- solve a system of (simultaneous) linear equations
- find the slope of a line
- understand the economic applications of linear equations
- understand the concept of supply and demand

In this chapter, we will explore all facets of linear equations. Keep in mind that linear equations are the most important concept in Algebra. Therefore it is recommended that you master this chapter before moving on to other chapters.
Students should be able to work with concepts such as *slope* and *y-intercept*.

Linear equations are part and parcel of our daily lives

Here are a few situations where we use linear equations:
When we travel: we calculate the distance and the speed of the traffic to estimate how late we will arrive at our destination.
When sharing things: we try to make sure that all parts are equally divided.
When shopping: we try to budget for the number of items we can purchase in relation to our income.
When we take a taxi: usually taxi drivers charge an initial *fixed fee* as part of using their services. Then, for each km travelled, they charge a certain amount on top.
A cell phone company usually charges us for the number of minutes we use in a month.

We just don't usually think of these in terms of equations and formula. We'd rather use language to describe such situations. But Math can be a great and easy means for translating those words into numbers. Linear equations allow us to solve complex and changeable problems using a set of simple equations. Constants and variables are combined in the equations.

Source: www.ehow.com

2.1 Equations and Linear Equations

Equations are certainly the most important tool in algebra.

Definition 2.1 Equation
An equation is a statement that has two equivalent quantities. The process of finding out the variable value that makes the equation true is called 'solving' the equation.

2

Example 2.1

An Example of an Equation is: $x+1=2$,
and the solution of this equation is: $x=1$.

Definition 2.2 Linear equation
A linear equation is an equation that is composed of two expressions set to be equal to each other. A linear equation should fulfill the following conditions:

- The equation has one or two variables.
- All variables of the equation are raised to the power of 1.
- The graph of the equation forms a straight line.

A linear equation can be written in the form:

$$y=ax+b$$

where:

- y is the depending variable of x
- x the independent variable
- a and b are constants
- x is always to the power of 1 (see figure 2.1)

FIGURE 2.1 The linear equation $y = ax + b$

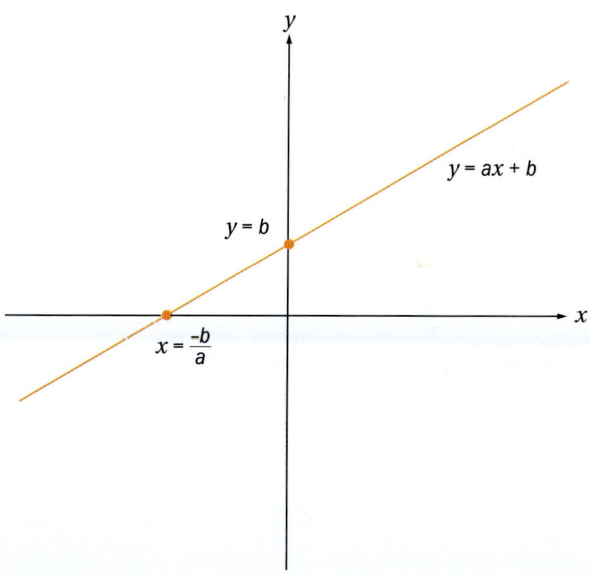

2.2 Linear Equations (LE) of One Variable

Example 2.2

Suppose we know that a variable x satisfies

$$3x+2=-2x+12 \tag{2.1}$$

and we want to find the value of x.

2

LE1 First we subtract 2 from both sides of the equation:
$3x=-2x+10$

LE1 Then we add $2x$ to both sides:
$5x=10$

LE2 Dividing both sides by 5 finally gives us the solution:

$$x=\frac{10}{5}=2 \tag{2.2}$$

THEOREM 2.1

Solving a Linear Equation

In solving the linear equation in the previous example we used the following properties: the validity of an equation doesn't change if:
LE1 any number is added to or subtracted from both sides of the equation
LE2 both sides of the equation are multiplied or divided by any number (not zero)

We can also check if our solution of (2.1) is correct; we insert the solution $x=2$ into the equation we had to solve:

$$\begin{aligned} 3x+2&=-2x+12\\ 3\times2+2&=-2\times2+12\\ 6+2&=-4+12\\ 8&=8 \end{aligned}$$

So the solution is correct.

2.3 The Equation of a Line

The equation of a line may appear in different forms, such as:

$$x+2y=2 \tag{2.3}$$

or

$$y=-\frac{1}{2}x+1 \tag{2.4}$$

Both of these equations describe the same line because they are equivalent. We demonstrate here how to derive equation (2.4) from (2.3):

$$x+2y=2$$
$$2y=-x+2 \text{ by subtracting } x \text{ (LE1)}$$
$$y=-\frac{1}{2}x+1 \text{ by dividing by 2 (LE2)}$$

Reversing these steps shows that equation (2.3) can also be derived from equation (2.4). Later in this chapter we will find that $-\frac{1}{2}$ is the slope of this line, and 1 the y-intercept.

2

2.4 Graphing a Linear Function

FIGURE 2.2 The line $y = -\frac{1}{2}x + 1$

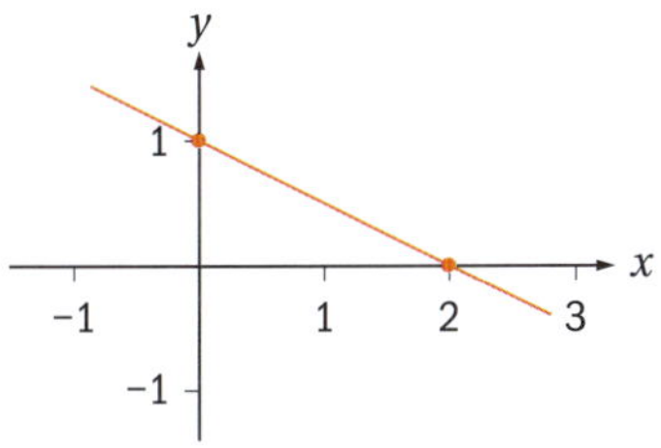

A straight line is determined between two different points. Think of spanning a string between two points; one point would not be enough, three would be too many. The same principle holds for drawing a graph of a linear function; only two different points are needed. Consider the following example:

We draw a graph of the line $y=-\frac{1}{2}x+1$. Inserting 0 for x yields $y=1$, hence the point (0,1) is on this line. First we put this coordinate in a table (see table 2.1). Another point can be found by inserting 0 for y and solving this equation for x:

$$0=-\frac{1}{2}x+1$$
$$-\frac{1}{2}x=-1$$
$$x=2$$

Hence another point on the line is (2, 0). We will also put this coordinate in the table. Now we can plot these points (0, 1) and (2, 0) in an (x, y) axes system, and draw a straight line between these points (figure 2.2).

TABLE 2.1

x	0	2
y	1	0

2.5 Slope of a Line

The slope tells us how steep a line is. An equation becomes a line when it is graphed (figure 2.3); you can tell from looking at the equation what the slope is.

- If the slope is positive, the line goes upwards.
- If the slope is negative, the line goes downwards.
- If the slope is zero, it's a horizontal line (see figure 2.3).

FIGURE 2.3 The slope of a line

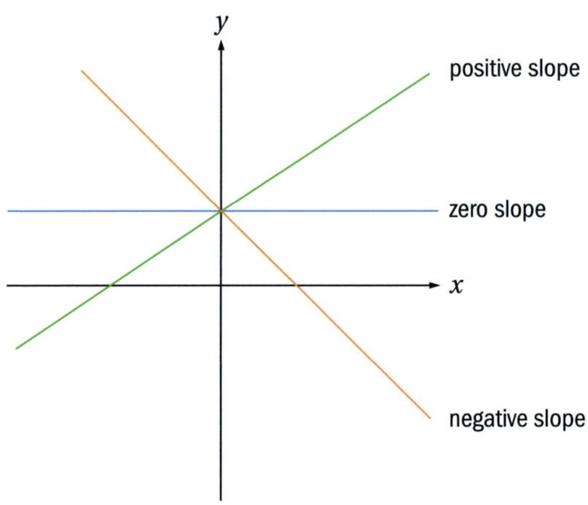

Definition 2.3 Slope of a Line

The slope of a line is the ratio of the change in y over the change in x, denoted by

$$\frac{\Delta y}{\Delta x}$$

The symbol Δ (Delta, the Greek letter D) stands for Δ Difference. We divide the difference of two y-values y_1 and y_2 by the difference of the two corresponding x-values x_1 and x_2:

$$\frac{\Delta y}{\Delta x} = \frac{y_2 - y_1}{x_2 - x_1} \tag{2.5}$$

In order to find the slope of a line, e.g. $y = -\frac{1}{2}x + 1$, we insert two different coordinates into this formula. Let us insert the coordinates found in table 2.1 into formula (2.5):

$$\frac{\Delta y}{\Delta x} = \frac{y_2 - y_1}{x_2 - x_1} = \frac{0-1}{2-0} = \frac{-1}{2}$$

Which means that if x increases by 1, y increases by $-\frac{1}{2}$, and if x increases by 2, y increases by -1. In this case the slope is negative, hence the line is descending. Find this feature in the graph of this line, figure 2.2.

Since inserting 0 for x into $y = -\frac{1}{2}x + 1$ yields the coordinate (0, 1), which is the y-intercept. We now have identified $-\frac{1}{2}$ as the slope, and 1 as the y-intercept of the line $y = -\frac{1}{2}x + 1$.

THEOREM 2.2

Slope and *y*-intercept of a Line

In the equation of a line:

$$y = ax + b$$

The slope of this line is a and the y-intercept of this line is b. For $a > 0$ the line is ascending and for $a < 0$ the line is descending.

For $a = 0$ the equation reduces to

$$y = b,$$

this is a horizontal line.

Applications of the Slope

1 The slope, a, in a real life situation could refer to a rate or frequency, for example:
- 15 liters/hour
- $ 5/person
- 10 kg /week

2 The y-axis intercept (b) is the flat fee or the fixed quantity, which is always constant.

So
- flat fee = b (y-intercept)
- rate = a (slope)

2

Example 2.3

Suppose you are taking a taxi to go to a concert. The taxi driver charges you a fixed fee of €4 for using its service. But for each km (distance) travelled, the taxi driver charges you an additional fee of €1.25.

1 Write the linear equation that could be used to find the total bill when your final destination is reached (Hint: let x be the number of km and the y the total cost).
2 How much would the taxi driver charge if you travelled 10 km?

Solution 1:

Flat fee	b (y-intercept)	= €4
Rate	a (slope)	= €1.25

$y = ax + b$

$y = 1.25x + 4$

How much would you pay the taxi driver if he drove 10 km? First notice that $x = 10$ and substitute 10 for x into the equation, hence

$$\begin{aligned} y &= 1.25x + 4 \\ &= 1.25 \times 10 + 4 \\ &= 16.50 \end{aligned}$$

So you would have to pay €16.50.

2.6 Economic Application of Linear Equations: Supply and Demand Models

The Cartesian Plane is very important for understanding the concepts of supply and demand, as these are the most fundamental concepts in economics.

> **Definition 2.4 Demand (D)**
> *Demand* refers to the quantity of a product or service that is desired by buyers. And *quantity demand* refers to the number of products or services a buyer is willing and is able to purchase at a specific point in time.

The quantity of goods or services that a consumer is willing to pay for depends on many factors. The most common are:

- their tastes and preferences
- the price of the goods
- the price of alternative and complementary goods
- the consumer's income
- the expectation of future price change

The law of demand states that:
When the price of certain goods rises, the quantity demanded falls (ceteris paribus). That is also why the demand curve always *slopes downwards*, in fact, and the quantity and price curve changes in a different direction. See figure 2.4.

FIGURE 2.4 The demand curve slopes downwards

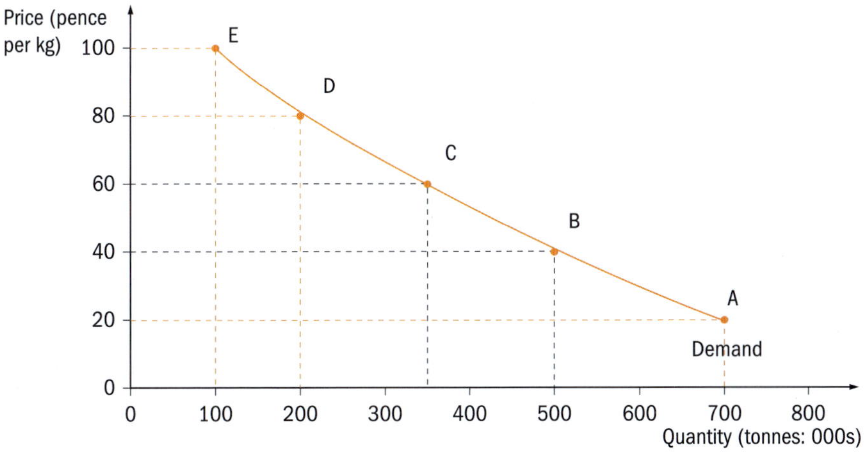

Source: Sloman, 5th edition

Definition 2.5 Demand Function or Equation
The demand function or equation expresses demand Q_d (the number of items demanded as a function of the unit price P (the price per item). The demand equation is given by:

$$Q_d = a - bP$$

Where Q_d is the quantity, P is the price and a and b are constants.

Definition 2.6 Supply
Supply represents how much the market can offer. The quantity supplied refers to the amount of a particular type of goods that producers are willing to supply against a certain price.

The law of supply states that:
When the price of goods rises, the quantity supplied also rises (ceteris paribus). That is why the supply curve always *slopes upwards*. In fact, quantity and price change in the same direction. See figure 2.5.

The quantity of goods or services that a producer is willing to sell depends on many factors. The most common are:
- the price of the goods
- the cost of production
- the price of alternative goods
- the number of suppliers
- changes in technology

FIGURE 2.5 The supply function slopes upwards

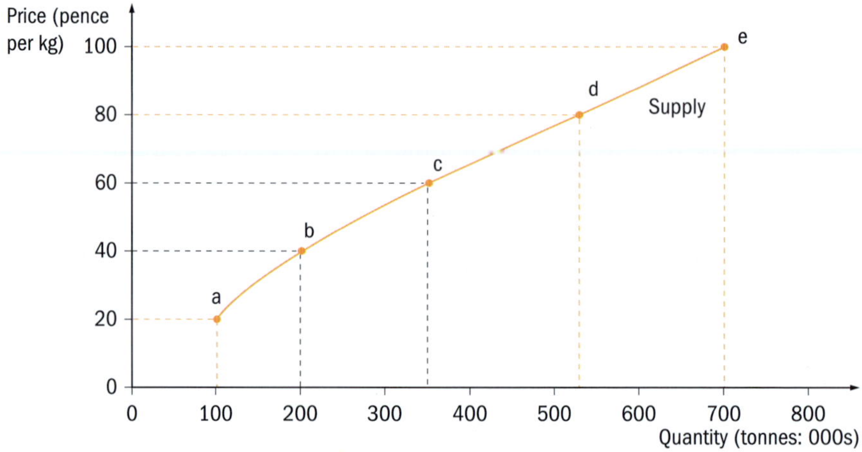

Source: Sloman, 5th edition

Definition 2.7 Supply Equation or Function
The supply equation or function expresses supply Q_s (the number of items the supplier is willing to bring to the market) as a function of the unit price P (the price per item):

$$Q_s = c + dP$$

Where Q_s is the quantity supplied, P is the price, and c and d are constants.

Note: Since supply slopes upwards, its slope is positive, and as demand slopes downwards, its slope is negative. This explains the difference in the equations for supply and demand.

Equilibrium Point
The equilibrium point is the point where the demand and supply curves intercept. In other words, it is the point where $Q_d = Q_s$ (see figure 2.6). The equilibrium creates two situations:
- a shortage (point below the equilibrium) where $Q_d > Q_s$
- a surplus (point above the equilibrium) where $Q_d < Q_s$

2.7 Simultaneous Linear Equations

Suppose your car breaks down, so you call a mechanic. Just for showing up this guy will charge € 30, and for every hour of work he charges an additional €40. This means that if the mechanic spends 1 hour fixing your car, the bill is €30 + 1 × €40 = €70. This is an example of a linear equation. In this case the equation is:

$$y = 40x + 30,$$

FIGURE 2.6 The equilibrium point of supply and demand

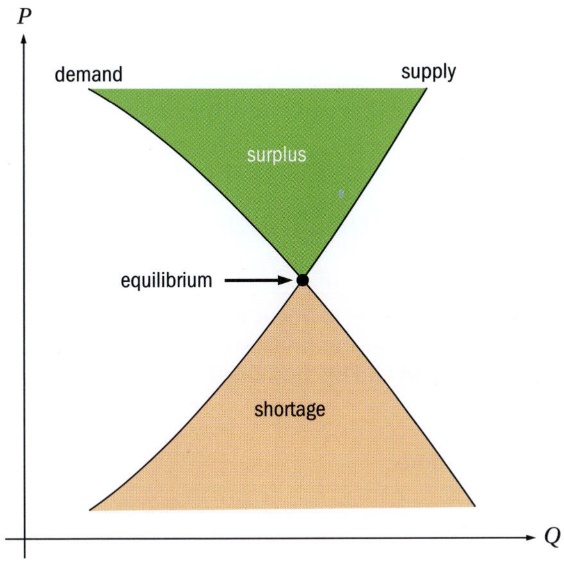

where y is the invoice, and x is the number of hours worked by the mechanic. You, however, as a financial advisor, charge €70 per hour. After one hour both you and the mechanic charge the same amount of money, €70. This is an example of the solution of a system of 2 equations:

$$\begin{cases} y = 30 + 40x \\ y = 70x \end{cases}$$

which has solution $x = 1$ and $y = 70$.

We will now explain how to solve a system of 2 equations.

A system of simultaneous linear equations, such as:

$$\begin{cases} x + 2y = 2 \\ 2x - 2y = 1 \end{cases} \tag{2.6}$$

is a set of 2 linear equations that are set out such that a solution to one of the equations is also a solution to the other. For example: $x = 1$ and $y = \frac{1}{2}$ is a solution of both of the equations:

$$\begin{cases} 1 + 2 \times \frac{1}{2} = 2 \\ 2 \times 1 - 2 \times \frac{1}{2} = 1 \end{cases}$$

As another example: $x = 0$ and $y = 1$ solves the first equation $x + 2y = 2$ since $0 + 2 \times 1 = 2$, but does not solve the other equation $2x - 2y = 1$, since

$2 \times 0 - 2 \times 1 = -2 \neq 1$. We call the solution $x = 1$ and $y = \frac{1}{2}$ the solution of the system of equations (2.6) because it solves both of the equations of this system.

There are several ways to solve a set of linear equations. A couple of them will be discussed in the following subsection.

2.8 Solving a Set of Linear Equations

How to solve a set of linear equations will be discussed in the following subsection. There are several ways to solve a set of linear equations graphically. In this paragraph we discuss three of those methods: solving equations graphically, by the elimination method and by the substitution method.

2.8.1 Solving Graphically

The solution of this system of equations:

$$\begin{cases} x + 2y = 2 \\ 2x - 2y = 1 \end{cases}$$

can be visualized by drawing a graph of both functions in one set of axes (see figure 2.7). For the first equation from the system, $x + 2y = 2$ we have for $x = 0$ that $y = 1$, and for $y = 0$ that $x = 2$, hence the points (0, 1) and (2, 0) determine the graph of the first equation. These coordinates are tabulated in table 2.2: $x + 2y = 2$, plotted on a graph and joined by a straight line. Now we have the first (descending) line. For the second equation, $2x - 2y = 1$, we have for $x = 0$ that $y = -\frac{1}{2}$ and for $y = 0$ that $x = \frac{1}{2}$. Hence the points $\left(0, -\frac{1}{2}\right)$ and $\left(\frac{1}{2}, 0\right)$ determine the graph of the second equation. Again, these coordinates are tabulated in table 2.3, plotted on a graph and joined by a straight line. Now we have the second (ascending) line. The point of intersection is the solution of the system of equations because the values of x and y satisfy both of the equations.
N.B. Graphical solving only provides an estimated solution.

TABLE 2.2 $x + 2y = 2$

x	0	2
y	1	0

TABLE 2.3 $2x - 2y = 1$

x	0	$\frac{1}{2}$
y	−1	0

FIGURE 2.7 The lines $x + 2y = 2$ and $2x - 2y = 1$

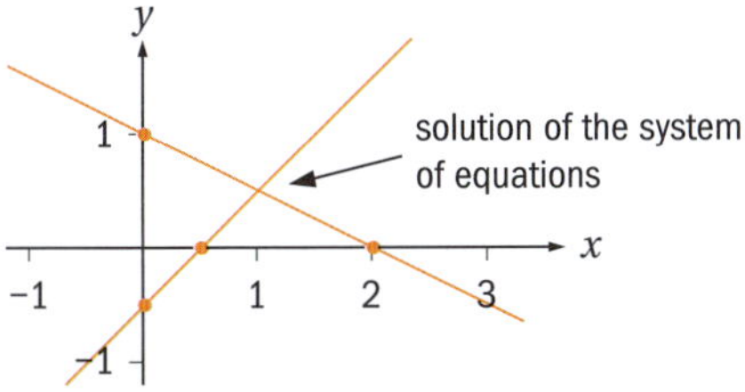

2.8.2 Solving by the Elimination Method

Consider the following set of simultaneous equations:

$$\begin{cases} x+2y=2 \\ 2x-2y=1 \end{cases} \tag{2.7}$$

The terms $2y$ and $-2y$ will be eliminated if added, so by adding the equations we can eliminate y:

x	$+$	$2y$	$=$	2	
$2x$	$-$	$2y$	$=$	1	$+$
$3x$			$=$	3	

Hence $x=1$. Now if we substitute this solution in either of the equations (2.7), we will find the value for y. For example, substituting 1 for x in the first equation, $x+2y=2$, gives us:

$$\begin{aligned} x+2y&=2 \\ 1+2y&=2 \\ 2y&=1 \\ y&=\frac{1}{2} \end{aligned}$$

So $x=1$ and $y=\frac{1}{2}$ solve the set of simultaneous equations (2.7). Substitution of $x=1$ in the second equation will result in $y=\frac{1}{2}$ too.

Another example:

$$\begin{cases} 2x+3y=1 \\ 3x+2y=4 \end{cases} \tag{2.8}$$

Adding or subtracting both equations brings us no closer to the solution. However, property LE2 tells us that the solutions of the set of the equations do not change if we multiply them by any number unequal to 0. Multiplication of the first equation by 3 and the second by 2 yields:

$$\begin{cases} 6x+9y=3 \\ 6x+4y=8 \end{cases}$$

Now we can eliminate x by subtracting the equations from one another:

$$\begin{array}{r} 6x+9y=3 \\ \underline{6x+4y=8-} \\ 5y=-5 \end{array}$$

Hence $y=-1$. Substitution of -1 for y in either of the equations (2.8) yields $x=2$.

2.8.3 Solving by the Substitution Method

Consider the following set of simultaneous equations:

$$\begin{cases} x+\ \ y=3 \\ 2x+3y=8 \end{cases} \tag{2.9}$$

We will solve this equation using the substitution method.

1 Isolate the variable x in the first equation:

$$x+y=3$$

Isolating x gives:

$$x=3-y$$

2 Substitute the isolated variable in the second equation:
Substituting $3-y$ yields:

$$2(3-y)+3y=8$$

This linear equation has only one variable, so we can solve it.

3 Solve this equation for the other variable, y:

$$2(3-y)+3y=8$$

Expanding the brackets:

$$6-2y+3y=8$$

Simplifying gives:

$$6+y=8$$

Hence $y=2$.

4 Substitute the known value of y into the equation for x derived in step 1:

$$x=3-y$$

Substituting 2 for y yields:

$x=3-2$
$x=1$

So $x=1$ and $y=2$ are the solutions for this system of equations (2.9).

THEOREM 2.3

Solving a System of 3 or More Equations

A system of 3 simultaneous equations such as:

$$\begin{cases} x+2y+z=1 \\ x+y-z=0 \\ 2x+y+z=4 \end{cases} \tag{2.10}$$

can be solved by eliminating z from the first two of the three equations:

$$\begin{array}{l} x+2y-z=1 \\ \underline{x+y-z\ =0\quad +} \\ 2x+3y\ =1 \end{array}$$

and eliminating z from the last two of the three equations:

$$\begin{array}{l} x+y-z\ =0 \\ \underline{2x+y+z=4\quad +} \\ 3x+2y\ =4 \end{array}$$

This yields the system of 2 simultaneous equations (2.8):

$$\begin{cases} 2x+3y=1 \\ 3x+2y=4 \end{cases}$$

which has solutions $x=2$ and $y=-1$ as found in the previous subsection. Inserting these solutions into either one of the equations from (2.9), e.g. the last, yields

$$\begin{aligned} 2x+y+z&=4 \\ 2\times 2-1+z&=4 \\ z&=1 \end{aligned}$$

Hence the solution of (2.9) is $x=2$, $y=-1$ and $z=1$.

2.9 The Equation of a Line Revisited

Just as we can find coordinates on a line if the equation of this line is given, we can also determine the equation of the line if two or more coordinates are given (see figure 2.8).

FIGURE 2.8 A line through (0, 1) and (2, 0)

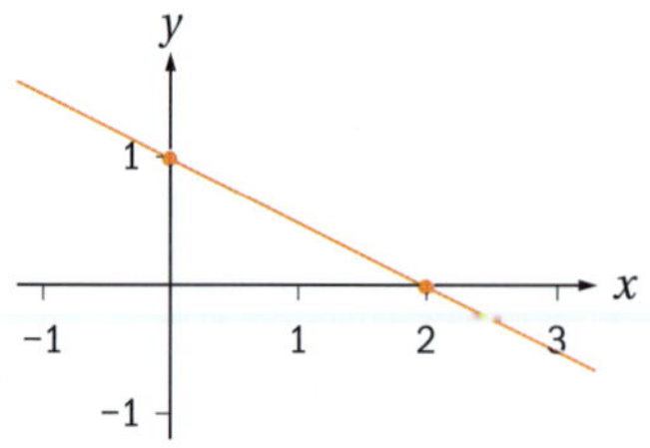

2

Example 2.4

Let us find the equation of a line through the points $(x_1, y_1)=(0, 1)$ and $(x_2, y_2)=(2, 0)$. We substitute the values for (x_1, y_1) into the general equation of a line $y=ax+b$ and find

$$1=a\times 0+b \tag{2.11}$$

And we do the same for (x_2, y_2):

$$0=a\times 2+b \tag{2.12}$$

These two equations (2.11 and 2.12) are the set of equations:

$$\begin{cases} b=1 \\ 2a+b=0 \end{cases}$$

This system has solutions $a=-\frac{1}{2}$ and $b=1$. Hence the equation of a line $y=ax+b$ through the points (0, 1) and (2, 0) is

$$y=-\frac{1}{2}x+1$$

This line has slope $-\frac{1}{2}$ and y-intercept 1.

Summary

In order to solve a linear equation one can:

LE1 add any number to or subtract any number from both sides of the equation

LE2 multiply or divide both sides of the equation with any number (not zero)

E.g.:

$$3x+2=x+5$$

– LE1: subtract 2 from both sides:

$$3x=x+3$$

– LE1: subtract x from both sides:

$$2x=3$$

– LE2: divide both sides by 2:

$$x=\frac{3}{2}$$

► Plotting a linear function is done by calculating the y-coordinate for $x=0$, then the x-coordinate for $y=0$, plotting both of these points on a graph and joining them by a straight line, e.g.:

– $y=-\frac{1}{2}x+1$

x	0	2
y	1	0

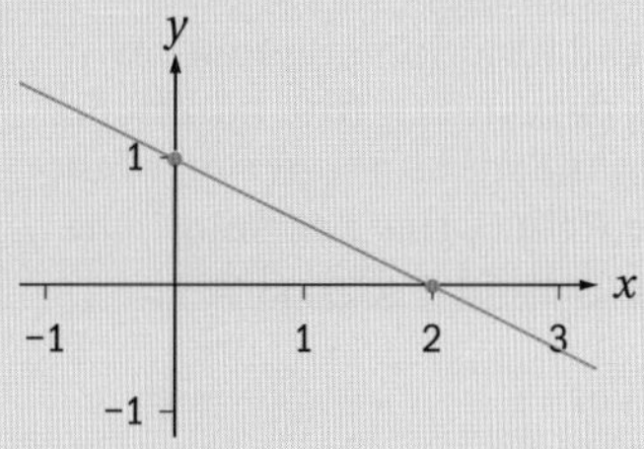

► In the general formula of a straight line:

$$y=ax+b$$

a is the slope, and b the y-intercept.

► A system of 2 or more equations, such as:

$$\begin{cases}2x+3y=1\\3x+2y=4\end{cases}$$

can be solved by multiplying the first of these equations by 3 (the coefficient of x of the second equation), and multiplying the second equation by 2 (the coefficient of x of the first equation), giving:

$$\begin{cases}6x+9y=3\\6x+4y=8\end{cases}$$

Now we can eliminate x by subtracting the equations from one another:

$$\begin{array}{r}6x+9y=3\\ \underline{6x+4y=8\;-}\\ 5y=-5\end{array}$$

Hence $y=-1$. Substitution of -1 for y in either of the equations:

$$\begin{cases} 2x+3y=1 \\ 3x+2y=4 \end{cases}$$

yields $x=2$. Hence the solution is $x=2$ and $y=-1$.

- Demand Function or Equation
  The demand function or equation expresses demand Q_d (the number of items demanded) as a function of the unit price P (the price per item). The demand equation is given by:

  $$Q_d = a - bP$$

  Where Q_d is the quantity, P the price, and a and b are constants.

- Supply Equation or Function
  The supply equation or function expresses supply Q_s (the number of items the supplier is willing to bring to the market) as a function of the unit price P (the price per item)

  $$Q_s = c + dP$$

  Where Q_s is the quantity supplied, P is the price, and c and d are constants.

- Equilibrium Point
  The equilibrium point is the point where demand and supply curves intercept. In other words, it is the point where $Q_d = Q_s$. The equilibrium creates two situations:

- a shortage (point below the equilibrium) where $Q_d > Q_s$

- a surplus (point above the equilibrium) where $Q_d < Q_s$

Exercises

Exercises 2.2, 2.11 and 2.12 are real life applications of linear equations.

2.1 Think of a number to which you would add 3 in order to give the result of 8. What is that number? Write the linear equation of this process.

2.2 During a school event, the organization is using an age-related pricing structure such that children's tickets cost €6 each and adult tickets cost €14 each. The total amount of money earned from ticket sales equals €420.

a Write a linear model that relates the number of children's tickets sold to the number of adult's tickets sold.

b How many children's tickets were sold if 15 adult tickets were sold?

2.3 Solve

a $7x = 14$

b $5x + 1 = 3x + 2$

c $2x + 2 = x - 2$

2.4 Plot the graph of $x + 2y = 2$. First calculate the y-intercept by inserting $x = 0$, then the x-intercept by inserting $y = 0$. Why is this the same graph as in figure 2.2?

2.5 Plot the graphs of

a $y = 2x + 1$

b $y = 1 - x$

c What are the coordinates of the point of intersection? This is the solution of the system of equations:

$$\begin{cases} y = x \\ y = 1 \end{cases}$$

2.6 Plot the graphs of

a $y = x$

b $y = 1$

c What are the coordinates of the point of intersection? This is the solution of the system of equations:

$$\begin{cases} y - 2x = 1 \\ x + y \ = 1 \end{cases}$$

2

2.7 Solve these sets of equations:

a $\begin{cases} x+y=3 \\ x-y=-1 \end{cases}$

b $\begin{cases} 2a+3b=13 \\ -a-3b=-14 \end{cases}$

c $\begin{cases} -2u+3v=2 \\ 4u-3v=-3 \end{cases}$

d $\begin{cases} 2x+3y=1 \\ 3x+2y=2 \end{cases}$

2.8 Solve these sets of equations:

a $\begin{cases} x+y+z=3 \\ x+y-z=1 \\ x-y-z=-1 \end{cases}$

b $\begin{cases} a+2b+3c=14 \\ 3a+2b+c=10 \\ 2a-2b+2c=4 \end{cases}$

c $\begin{cases} 2u+3v+4w=3 \\ 4u-3v-4w=0 \\ 12u-12v-12w=-1 \end{cases}$

d $\begin{cases} 2x+3y+z=2 \\ 3x+2y-z=1 \\ x-y-z=0 \end{cases}$

e $\begin{cases} u+v=2 \\ v+w=0 \\ u+w=1 \end{cases}$

2.9 Find the equations of the lines through:

a (1, 4) and (2, 7)
b (5, −6) and (1, −2)
c (0, 1) and (1, 1)
d What are the slopes and y-intercepts of these lines?

2.10 Label the boxes with the following:

- Which axis represents the price?
- Which axis represents the quantity?
- Indicate which line represents the demand?
- Indicate which line represents the supply?
- Indicate the equilibrium point.

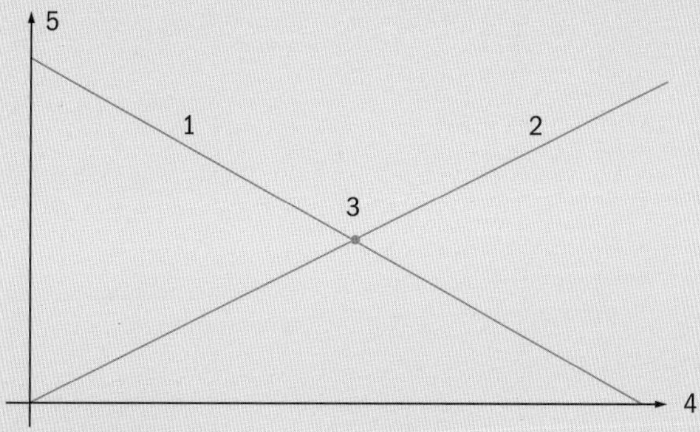

2.11 The table below represents the Quantity demanded and the Quantity supplied of a company making cake:

Price of a piece of cake	Quantity demanded	Quantity supplied
€1.20	3,000	1,200
€1.30	2,500	1,400
€1.60	2,400	1,800
€1.80	2,000	2,000
€2.00	1,600	2,200
€2.20	1,200	2,400

a Make a graph representing the curves of demand and supply.
b Indicate the equilibrium point.

2.12 The demand equation and the supply equation of a certain company are given by:
$Q_d = -100p + 300$
$Q_s = 120p - 80$

a Calculate the equilibrium point.
b Draw the supply curve and the demand curve on a graph.

3 Non-linear Functions and Equations

After studying this chapter you should be able to:

- solve quadratic equations
- graph quadratic equations
- understand the basic manipulations of powers
- solve equations involving exponents and logarithms

Now that you are familiar with the concept of equations, we will move on to a more complex form of equations: non linear equations. Recall that in chapter 2, we stated that the standard formula of a linear equation is $y = ax + b$, and the graph is a straight line. In this chapter we will discuss non-linear equations by focusing more on quadratic equations. We will also discuss how to identify the roots of a quadratic equation which allow us to investigate the application of non linear equations in major business, scientific and economic problems.

Quadratic equations in ancient times

From around 800 B.C. quadratic equations were being studied in India. Later Euclid and Pythagoras also tried to find their solutions. But it was not until the Persian mathematician and astronomer Ab ū 'Abdallāh Muhammad ibn Mūsā al-Khwarizmi wrote the book 'Kitab al-Jabr wa-l-Muqabala' (after which the discipline algebra, al-Jabr, was named; also compare Muqabala with Kabbalah, the Jewish mystic book of numbers), in around 820 A.D., that a systematic way of solving quadratic equations was found.

3.1 Quadratic Equations

A quadratic equation is an equation which can be written in the form:

$$y = ax^2 + bx + c$$

FIGURE 3.1 The quadratic equation

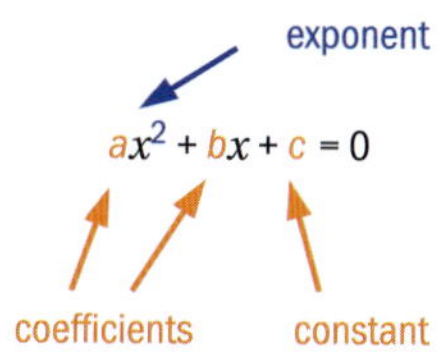

or with an example:

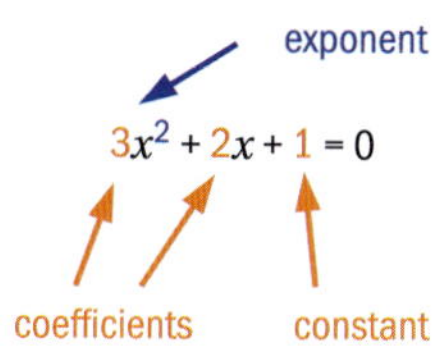

In this equation a and b are coefficients and c is a constant. This equation is called quadratic because of the exponent 2. The first term of a quadratic equation always is ax^2, with a not equal to 0.

3.1.1 How to Solve a Quadratic Equation

FIGURE 3.2 The parabola $y = x^2 - x - 2$

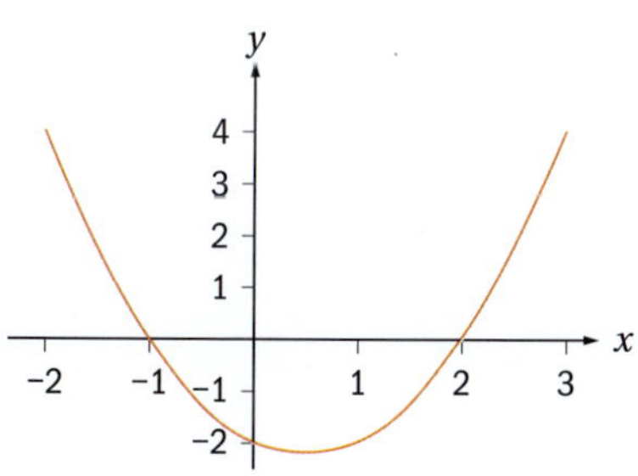

Definition 3.1 Perfect Square
Consider the expression $(a+b)^2$:

$$(a+b)^2 = (a+b)(a+b)$$

Expanding brackets gives us:

$$\begin{aligned} &= a(a+b) + b(a+b) \\ &= a^2 + ab + ba + b^2 \\ &= a^2 + 2ab + b^2 \end{aligned}$$

Thus $a^2+2ab+b^2=(a+b)^2$

Now consider the expression $(a-b)^2$:

$$(a-b)^2=(a-b)(a-b)$$

Expanding brackets gives us:

$$\begin{aligned}&=a(a-b)-b(a-b)\\&=a^2-ab-ba+b^2\\&=a^2-2ab+b^2\end{aligned}$$

Thus $a^2-2ab+b^2=(a-b)^2$

Now consider the expression $(a+b)(a-b)$:

$$\begin{aligned}(a+b)(a-b)&=a(a-b)+b(a-b)\\&=a^2-ab+ba-b^2\\&=a^2-b^2\end{aligned}$$

Thus $a^2-b^2=(a+b)(a-b)$

Example 3.1 Factorize the Following Expressions

$$\begin{aligned}(x+5)^2&=(x+5)(x+5)\\&=x^2+5x+5x+5^2\\&=x^2+10x+25\end{aligned}$$

Thus $x^2+10x+25=(x+5)^2$

$$\begin{aligned}(x-3)^2&=(x-3)(x-3)\\&=x^2-3x-3x+3^2\\&=x^2-6x+9\end{aligned}$$

Thus $x^2-6x+9=(x-3)^2$

$$\begin{aligned}&x^2-16\\&\quad=x^2-4^2\\&\quad=(x-4)(x+4)\end{aligned}$$

Thus $x^2-16=(x-4)(x+4)$

3.2 Solving a Quadratic Equation Using the Perfect Square

In this paragraph, we solve a quadratic equation using the perfect square.

$$\begin{aligned}&x^2-1=0\\&\quad\Rightarrow(x-1)(x+1)=0\\&x=1\\&x=-1\end{aligned}$$

$$\begin{aligned} x^2+6x+9&=0 \\ \Rightarrow (x+3)(x+3)&=0 \\ x&=-3 \end{aligned}$$

$$\begin{aligned} x^2-8x+16&=0 \\ \Rightarrow (x-4)(x-4)&=0 \\ x&=-4 \end{aligned}$$

THEOREM 3.1

The *abc*-formula

Where $a \neq 0$, and a, b, and c work out such that $b^2-4ac \geq 0$ provides the solutions x_1 and x_2 of the quadratic equation

$$ax^2+bx+c=0 \tag{3.1}$$

therefore:

$$x_1=\frac{-b-\sqrt{b^2-4ac}}{2a} \text{ and } \qquad x_2=\frac{-b+\sqrt{b^2-4ac}}{2a}. \tag{3.2}$$

The term $D=b^2-4ac$ is called the Discriminant.
The Discriminant quickly tells you the number of real roots, or in other words, the number of x-intercepts, associated with a quadratic equation:

- If $D=0$ there is one solution (or two identical solutions).
- If $D>0$ there are 2 solutions.
- If $D<0$ there are no solutions (see figure 3.3).

FIGURE 3.3 The Discriminant $D = b^2 - 4ac$ vs. the number of solutions of a quadratic equation

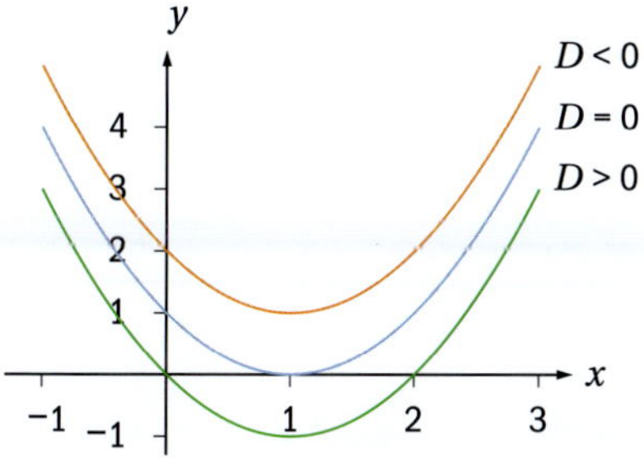

EXTEND YOUR KNOWLEDGE

Division of the example in 3.1 by a followed by a subtraction of $\frac{c}{a}$ from both sides of the equation gives $x^2+\frac{b}{a}x=-\frac{c}{a}$ (*). Now note that

$$\left(x+\frac{b}{2a}\right)^2-\frac{b^2}{4a^2}=\left(x^2+\frac{b}{a}x+\frac{h^2}{4a^2}\right)-\frac{h^2}{4a^2}=x^2+\frac{h}{a}x,$$

which is equal to the left hand side of (*). So

$$\left(x+\frac{b}{2a}\right)^2-\frac{b^2}{4a^2}=-\frac{c}{a}, \text{ and}$$

$$\left(x+\frac{b}{2a}\right)^2=\frac{b^2}{4a^2}-\frac{c}{a}=\frac{b^2}{4a^2}-\frac{4ac}{4a^2}=\frac{b^2-4ac}{4a^2}.$$

Taking the positive and negative of the square root of both sides, remembering that $(-u)^2=u^2$ yields

$$x+\frac{b}{2a}=\pm\sqrt{\frac{b^2-4ac}{4a^2}}=\frac{\pm\sqrt{b^2-4ac}}{2a}, \text{ hence } x=\frac{-b\pm\sqrt{b^2-4ac}}{2a}$$

QED

Example 3.2

Let us give an example of how to solve the quadratic equation using the Discriminant (*abc*-formula)

$$x^2-x-2=0 \qquad (3.3)$$

Note that a graph of the function $y=x^2-x-2$ is given in figure 3.1. Hence we are looking for the x-intercepts of this function. From the graph, we see that these are at $x=-1$ and $x=2$. To solve the equation (3.3), we must first compare it with the general formula (3.2):

$$ax^2+bx+c=0$$
$$x^2-x-2=0$$

We recognize that $a=1$, $b=-1$, and $c=-2$. Now we substitute these numbers into the *abc*-formula:

$$x_1=\frac{-b-\sqrt{b^2-4ac}}{2a}=\frac{1-\sqrt{1-4\times1\times(-2)}}{2}=\frac{1-\sqrt{9}}{2}=\frac{-2}{2}=-1,$$

$$x_2=\frac{-b+\sqrt{b^2-4ac}}{2a}=\frac{1+\sqrt{1-4\times1\times(-2)}}{2}=\frac{1+\sqrt{9}}{2}=\frac{4}{2}=2.$$

Hence $x^2-x-2=0$ for $x=-1$ and $x=2$. In order to check the validity of our solutions we plug them into the equation: for $x=2$ we find

$2^2-2-2=0$, and for $x=-1$ we find $(-1)^2-(-1)-2=0$. We have therefore made no error in our calculation.

Let us give another example of how to solve the quadratic equation:

$$x^2+x=2x+2 \tag{3.4}$$

We cannot apply formula (3.2) directly here because this equation is in a different form. So we work our quadratic equation (3.4) into the standard form (3.2): $ax^2+bx+c=0$. So we know we need to first subtract 2 from both sides of the equation (3.4):

$$x^2+x-2=2x$$

And second we subtract the term $2x$ from both sides of the last equation:

$$x^2-x-2=0.$$

Now we can apply the *abc*-formula. This equation is the same as equation (3.3) and is already solved in the previous example.

3.3 Graphing the Quadratic

The quadratic $y=ax^2+bx+c$ can be plotted on a graph. We must first calculate the x-intercepts, and some other values. This information is then tabulated, and plotted in a graph (see figure 3.4).

We plot of the quadratic function $y=x^2-x-2$. In Example 3.2 we already calculated the coordinates of the x-intercepts: $(-1, 0)$ and $(2, 0)$. Some values for x which are smaller or bigger than the intercepts are $x=-2$ and $x=3$. We therefore need to tabulate the x-values from -2 to 3, and their corresponding y-values:

x	−2	−1	0	1	2	3
y	−4	0	−2	−2	0	4

FIGURE 3.4 First plot the points from the table, then join them by a curved line

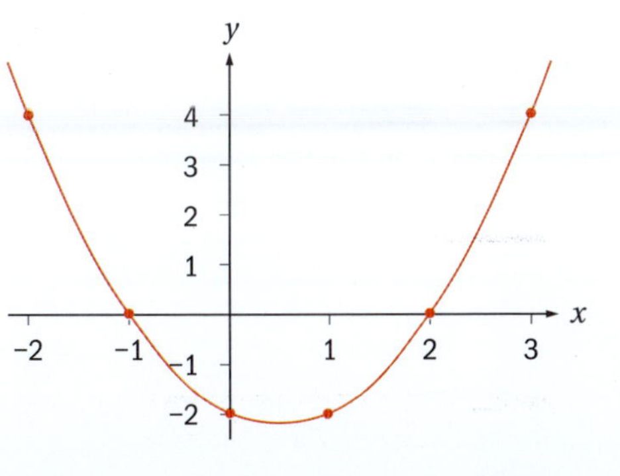

3.4 Powers

Just as multiplication is a shorthand for writing a repeated addition: $3+3+3+3=4\times3$, a power is a shorthand for writing a repeated multiplication: $3\times3\times3\times3=3^4$.

Definition 3.2 x to the power of n, x^n
For every x not equal to zero and every positive whole number n, x to the power of n, x^n, is defined by:

$$x^n = \underbrace{x \times x \times \ldots \times x}_{n \text{ times}}$$

$$x^0 = 1$$

$$x^{-n} = \frac{1}{x^n}$$

We call the number n the exponent of x.

Basically x^n is n number of x s multiplied by themselves.

This definition can be exemplified in table 3.1.

TABLE 3.1

Exponent	3^2				3×3	=	9	Number
	3^1	=	3	=	3	=	3	$\uparrow\times3$
$\uparrow+1$	3^0	=	1	=	1	=	1	
	3^{-1}	=	$\frac{1}{3}$	=	$\frac{1}{3}$	=	$\frac{1}{3}$	$\downarrow\div3$
$\downarrow-1$	3^{-2}	=	$\frac{1}{3^2}$	=	$\frac{1}{3\times3}$	=	$\frac{1}{9}$	

In table 3.1 we can see that the result is 3 times bigger when the exponent is increased by 1, or, when the number is divided by 3, the exponent decreases by 1.
Some rules apply when calculating using powers, these are as follows:

THEOREM 3.2

Calculation Rules for Powers

Let x and y be any numbers other than zero. Then for any n and m the following equalities apply:

$$x^n x^m = x^{n+m} \qquad (3.5)$$

$$\frac{x^n}{x^m} = x^{n-m} \tag{3.6}$$

$$(x^n)^m = x^{nm} \tag{3.7}$$

$$(xy)^n = x^n y^n \tag{3.8}$$

We will show how theorem 3.2 works in the following examples:

Example 3.3 Example of equation (3.5):

$$\begin{aligned} 2^3 \times 2^4 &= \underbrace{2\times2\times2}_{3\text{ times}} \times \underbrace{2\times2\times2\times2}_{4\text{ times}} \\ &= \underbrace{2\times2\times2\times2\times2\times2\times2}_{3+4\text{ times}} \\ &= 2^{3+4} \\ &= 2^7. \end{aligned}$$

Example 3.4 Example of Equation (3.7):

$$\begin{aligned} (2^3)^4 &= \underbrace{2^3\times2^3\times2^3\times2^3}_{4\text{ times}} \\ &= \underbrace{\underbrace{2\times2\times2}_{3\text{ times}}\times\underbrace{2\times2\times2}_{3\text{ times}}\times\underbrace{2\times2\times2}_{3\text{ times}}\times\underbrace{2\times2\times2}_{3\text{ times}}}_{4\text{ times}} \\ &= \underbrace{2\times2\times2\times2\times2\times2\times2\times2\times2\times2\times2\times2}_{3\times4\text{ times}} \\ &= 2^{3\times4} \\ &= 2^{12} \end{aligned}$$

3.5 Roots

The inverse operation of taking the n th power is taking the n th root. I.e. the solution of

$$x^3 = 8 \tag{3.9}$$

is

$$x = \sqrt[3]{8} = 2$$

because $2^3 = 8$.

FIGURE 3.5 $\sqrt[3]{8} = 2$ because $2^3 = 8$

Definition 3.3 *n*th root

Let $x \geq 0$ and n be a positive whole number. The nth root of x,

$$y = \sqrt[n]{x}$$

is defined to be that number y which, when raised to the power n equals x, i.e.

$$y^n = (\sqrt[n]{x})^n = x \tag{3.10}$$

Compare this definition of the nth root with Definition 3.2.
The square root of x, $\sqrt[2]{x}$, in general is written as $\sqrt{x}$.

Example 3.5

$$\sqrt[5]{32} = 2$$

because
$2^5 = 2 \times 2 \times 2 \times 2 \times 2 = 8$

THEOREM 3.3

$$\sqrt[n]{x} = x^{\frac{1}{n}}$$

For $x \geq 0$ and n, a positive whole number, we have

$$\sqrt[n]{x} = x^{\frac{1}{n}}$$

Explanation: According to equation (3.10) we have

$$(\sqrt[n]{x})^n = x \tag{3.11}$$

And according to equation (3.27) in Theorem 3.7 the left side of the last equation, Equation (3.11) raised to the power of $\frac{1}{n}$ equals:

$$((\sqrt[n]{x})^n)^{\frac{1}{n}} = (\sqrt[n]{x})^{\frac{n}{n}} = \sqrt[n]{x} \tag{3.12}$$

Whereas the right hand side of Equation (3.11) raised to the power of $\frac{1}{n}$ equals:

$$x^{\frac{1}{n}} \tag{3.13}$$

From Equation (3.13) and Equation (3.12) we have

$$x^{\frac{1}{n}} = \sqrt[n]{x}$$

3.6 Exponential Functions

Exponential growth occurs when the growth rate is equal to a certain percentage of the whole object that is growing (see figure 3.6). Important examples include the growth of the world's population: the number of babies born each year (the growth rate) equals a certain percentage of the world population (size of the growing object) in that given year. Another example is the growth of the balance of a bank account due to interest. If the interest rate is, say, 5% then the balance (size of the growing object) increases by 5% of that balance (growth rate) per year. The last example will be studied extensively in chapter 9.

FIGURE 3.6 Plots of various exponential figures

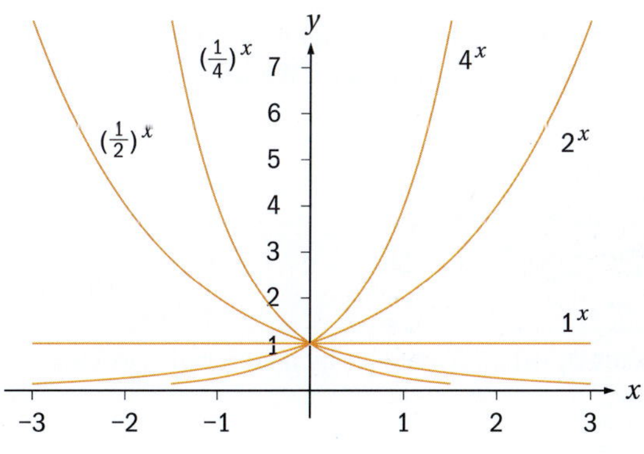

Definition 3.4 Exponential Function
For $a > 0$ the exponential function $y = f(x)$ with base a and variable x is defined by:

$$y = a^x \tag{3.14}$$

Note the similarities with the definition of the exponential function to the definition of powers, Definition 3.2. Also the calculation rules are similar to those of powers:

THEOREM 3.4

Calculation Rules for Exponential Functions

Let a and b be any positive numbers. Then for any x and y the following equalities apply:

$$a^0 = 1 \tag{3.15}$$

$$a^1 = a \tag{3.16}$$

$$a^{-x} = \left(\frac{1}{a}\right)^x = \frac{1}{a^x} \tag{3.17}$$

$$a^x a^y = a^{x+y} \tag{3.18}$$

$$\frac{a^x}{a^y} = x^{x-y} \tag{3.19}$$

$$(a^x)^y = a^{xy} \tag{3.20}$$

$$(ab)^x = a^x b^x \tag{3.21}$$

Note that for $a > 1$ the function a^x is increasing, and for $0 < a < 1$ the function a^x is decreasing, as can be seen in figure 3.6. A special exponential function is the exponential function e^x with base $e = 2.71828...$ We will meet e^x again in chapter 4 on differentiation.

Example 3.6

The exponential function

$$y = 5^x \tag{3.22}$$

has base $a = 5$. If x increases by one, y increases by a factor of 5, as can be seen in the table below.

x	−2		−1		0	1	2
y		$\frac{1}{25}$		$\frac{1}{5}$	1	5	25

The difference between a power function and an exponential function is where x is found:

power	x^5
exponential	5^x

3.7 Logarithms

Logarithms were invented by Napier in 1614. The logarithmic function is the inverse of an exponential function; hence logarithms are useful for solving an equation like:

$$5^x = 625 \tag{3.23}$$

Compare this equation with equation (3.22) and (3.9). Logarithms can be used to calculate how long it would take for an amount of money in the bank to reach a certain value. We will explain what logarithms are by answering the following question: 'To what power do you have to raise 2 in order to get 8?' The answer is 3 because

$$2^3 = 8$$

The same question: 'How much is

$\log_2(8)$?'

The answer is 3 because

$$2^3 = 8$$

As a memory aid:

$$\underset{\text{base}}{\log_{\downarrow 2}}(8) = \underset{\text{exponent}}{\overset{}{\downarrow 3}}$$

log $_2$(8) = 3, with an arrow from 2 to "base" and an arrow from 3 to "exponent".

Hence the logarithm of a with base b, the b logarithm of a, $\log_b(a)$, is just shorthand for 'to what power do you have to raise b in order to get a?'

THEOREM 3.5

Logarithm

Let $x > 0$ and $b > 0$, $b \neq 1$. The logarithm of x with base b,

$$y = \log_b(x),$$

allows us to find the value of x:

$$b^y = x$$

In other words, the solution of the equation $b^y = x$ is $y = \log_b(x)$.
We calculate

$$\log_3(9)$$

Since $3^2 = 9$ it follows that

$$\log_3(9) = 2$$

In practice $\log_{10}$ and ln are most often used. The ln is called the natural logarithm and is the logarithm with base e = 2.71828 . . . , so $\ln x = \log_e(x)$. The log button on your calculator returns the $\log_{10}$. In this book we adopt this convention.

THEOREM 3.6

Calculation Rules for Logarithms

Let x and y be any positive numbers. For any n the following qualities apply:

$$\log_b(xy) = \log_b(x) + \log_b(y) \tag{3.24}$$

$$\log_b\left(\frac{x}{y}\right) = \log_b(x) - \log_b(y) \tag{3.25}$$

$$\log_b(x^n) = n \times \log_b(x) \tag{3.26}$$

$$\log_c(x) = \frac{\log_b(x)}{\log_b(c)} \tag{3.27}$$

Example 3.7

We solve equation (3.23) by applying formula (3.26):
$5^x = 625$

Taking logarithms on both sides of the equation gives us:
$\log 5^x = \log 625$

Applying formula (3.26) gives us:
$x \log 5 = \log 625$

Now, dividing both sides of the equation by log 5 yields:

$$x = \frac{\log 625}{\log 5}$$

And finally, using your calculator:
$x = 4$

Let us solve the equation:
$11^x = 7 \times 13^x$

Taking logarithms from both sides yields:
$\log 11^x = \log(7 \times 13^x)$

Applying formula (3.24) on the right hand side of this equation gives us:
$\log 11^x = \log 7 + \log 13^x$

Now applying formula (3.26) gives us:
$x \log 11 = \log 7 + x \log 13$

Solving for x is done thus:
$x \log 11 - x \log 13 = \log 7$
$x(\log 11 - \log 13) = \log 7$

$$x = \frac{\log 7}{\log 11 - \log 13}$$

$$x \approx -11.65$$

Here is an application of formula (3.27): since the logarithms with any base b can be written as

$$\log_b(x) = \frac{\log_{10}(x)}{\log_{10}(b)} = \frac{1}{\log_{10}(b)} \times \log_{10}(x),$$

we can see that all logarithms with a different base are equal, up to a constant $\frac{1}{\log_{10}(b)}$.

Summary

- In order to solve a quadratic equation, we first need to work it into the standard form $ax^2 + bx + c = 0$, and apply the *abc*-formula:

$$x = \frac{-b \pm \sqrt{b^2 - 4ac}}{2a}$$

- A graph of a quadratic function is obtained by first tabulating some of its coordinates relating to the x-intercepts, then plotting these points and joining them by a curved line.

- x^n means n x-s multiplied by themselves:

$$x^n = \underbrace{x \times x \times \ldots \times x}_{n \text{ times}}$$

$$x^0 = 1$$

$$x^{-n} = \frac{1}{x^n}$$

- For powers the following rules apply:

$$x^n x^m = x^{n+m}$$

$$\frac{x^n}{x^m} = x^{n-m}$$

$$(x^n)^m = x^{nm}$$

$$(xy)^n = x^n y^n$$

- $\sqrt[n]{x}$ is that number which, raised to the power n equals x:

$$(\sqrt[n]{x})^n = x$$

- $\sqrt[n]{x} = x^{\frac{1}{n}}$

- For exponential functions $f(x) = a^x$ the following rules apply:

$$a^0 = 1$$

$$a^1 = a$$

$$a^{-x} = \left(\frac{1}{a}\right)^x$$

$$a^x a^y = a^{x+y}$$

$$\frac{a^x}{a^y} = x^{x-y}$$

$$(a^x)^y = a^{xy}$$

$$(ab)^x = a^x b^x$$

- The difference between a power function and an exponential function is where x is found:

power	x^5
exponential	5^x

- For $b > 0$, not equal to 1, and positive x, $y = log_b(x)$, is the solution of $b^y = x$.

- For logarithms the following rules apply:

$$\log xy = \log x + \log y$$

$$\log \frac{x}{y} = \log x - \log y$$

$$\log x^n = n \log x$$

$$\log_b(x) = \frac{\log_{10}(x)}{\log_{10}(b)}$$

Exercises

3.1 Factorize the following expressions:

a $3x^2+6x+3$
b $25-10y+y^2$
c $36-4u^2$
d $4x^2-y^2$
e $9x^2-1$
f $3x^2+6x+3$

3.2 Solve:

a $x^2-x=0$
b $x^2=x$
c $x^2-3x+2=0$
d $x^2-13x-7=0$
e $-x^2+11x-7=0$
f $x^2-11x+7=0$

3.3 Solve the following equations:

a $x^2-5x=3-3x$
b $2x^2-6x+4=x^2+5x-3$
c $x^2+2=5x-4$
d $2x^2-3x+5=x^2+4x-7$

3.4 Solve the following equations using the perfect square method:

a $x^2+10x+25=0$
b $x^2-6x+9=0$
c $x^2-16=0$
d x^2+x-2
e x^2-x-2
f x^2+x-6

3.5 Plot the graphs of x^2-4 and x^2-2x-3.

3.6* Show that the maximum or minimum of the quadratic ax^2+bx+c is located at $\frac{-b}{2a}$.

3.7 Show, analogue to equation 3.8 hat:
$(2\times 3)^4=2^4\times 3^4$

3.8 Show, analogue to equation 3.6 that:

$$\frac{2^3}{2^5}=\frac{1}{4}$$

3.9 Calculate:

a $\dfrac{2^{333}}{2^{332}}$

b $2^{333} \times 2^{-332}$

c $\dfrac{2^4}{4^2}$

d $\sqrt[3]{2^9}$

e $\dfrac{\sqrt{5^{200}}}{5^{51} \times 5^{50}}$

f $\sqrt[105]{3^{100} \times 3^{110}}$

g $\sqrt[3]{17 \times 17 \times 17}$

3.10 Simplify the following expressions by expanding the brackets and collecting like terms:

a $(x-y)(x^2+xy+y^2)$

b $(x-y)(x^3+x^2y+xy^2+y^2)$

c $(3x^3+3x^2-2x+5)-2x^3-2x^2+3x-4$

d $(x^3-5x^2-2x)-(-2x^3-6x-3x+11)$

3.11a Plot in one graph, for x ranging between -2 and 2, the following exponential functions:

$$3^x,\ 2^x,\ 1^x,\ \left(\frac{1}{2}\right)^x,\ \left(\frac{1}{3}\right)^x.$$

b Where do these functions intersect, and why do they do so there?

c Why doesn't 1^x increase or decrease?

3.12 Solve:

a $2^x = 4$

b $3^x = 4$

c $2^{x+1} = 3$

d $3^x = 4^x$

e $2 \times 3^x = 5$

f $2 \times 3^x = 5^x$

g $(2 \times 3)^x = 5$

h $2^x = 0.2$

3.13 Solve:

a $\left(\frac{1}{2}\right)^x = 4$

b $\frac{5^x}{7} = 3$

c $\left(\frac{5}{7}\right)^x = 3$

d $\left(\frac{5}{7}\right)^x = \frac{1}{2}$

e $\left(\frac{1}{2}\right)^x = 4$

f $\frac{3^x}{2} = 1$

g $\left(\frac{3}{2}\right)^x = 1$

h $\left(\frac{1}{2}\right)^x = 512$

3.14*

a How much is

$$\left((\sqrt{2})^{\sqrt{2}}\right)^{\sqrt{2}} ?$$

b Show that:

$$log_a(b) = \frac{1}{log_b(a)}$$

3.15*

The number of tulips in a field north east of The Hague is known to double every month. Furthermore it is known that it takes 24 months for this field to be full of tulips. How long does it take for this field to become half full?

4 Functions and Differentiation

At the end of this chapter, you should be able to:

- understand the concept of function.
- Find a slope or gradient of a function.
- Find the derivative of a function.
- find the maxima or the minima of a function.
- draw the graph of a function using specific points.
- understand the concept of price elasticity of demand.

Recall in chapter 2, we discussed the concept of 'slope of a line at a point' as the limit of secant slopes. In this chapter we will call that limit a derivative, which measures the rate at which a function changes.
The derivative plays an important role in Economics, Business and Science, for example it is used to calculate velocity and acceleration, and to estimate the rate of population growth. In business it helps in calculating and setting productivity levels and efficiency.
Business students should master this chapter, as it is the key to understanding chapter 5.

The origin of Differentiation

4

Differentiation was conceived in the 17th century by both Isaac Newton and Gottfried Leibniz, independently. At that time, neither man knew the other had come to the same conclusion, but when they found out they quarreled the rest of their lives about who was first.

Differentiation is extremely useful because it quantifies the rate of change of a function, that is, how fast it changes. When the rate of change is zero, the function is either at a maximum or a minimum. Think of jumping on a trampoline: the moments of time when you have no speed are when you are at the lowest and the highest points of the jump. For a price function these are the moments when you want to buy or sell, because the price is at a minimum or maximum.

4.1 Functions

We have now reached the most important chapter in algebra. In fact functions are so important that they are the heart of Math.

Definition 4.1 Function

FIGURE 4.1 A function is a set of mathematical operations with one or more variables

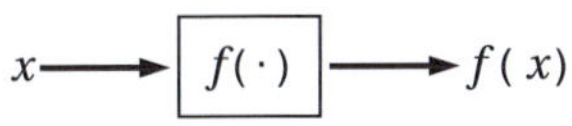

A function is a set of mathematical operations performed on one or more variables, that results in an output, e.g. $f(x) = x^2$:

$$\underbrace{f}_{\text{function name}}(\underbrace{x}_{\text{input}}) = \underbrace{x^2}_{\text{output}}$$

FIGURE 4.2 A banana tree

Example 4.1

This banana tree (figure 4.2) grows 40 cm every year, so the height of the banana tree is related to its age by the function *f*:
$f(age) = 40 \times age$
for $x = $ age
$f(x) = 40x$

So, if the age is 20 years (so $x = 20$), the height is:
$f(20) = 40x = 40 \times 20 = 800$ cm

4.2 Definition of the Derivative

The derivative of a function at a point is its instantaneous rate of change at that point. It is also the slope of the tangent line to the graph of the function at that point (provided that the tangent line exists and is not vertical). Given a function, $f(t)$ we can use the notation $f(t)$, to denote the derivative of f with respect to t at the point $t = a$ or the instantaneous rate of change of f at $t = a$. This concept will be illustrated in the following example, which is derived from http://earthmath.kennesaw.edu.

Suppose you are driving an accelerating car. Suppose that the distance s travelled at time t is given by $s(t) = t^2$. The question we will be answering is the following: what velocity does this car have at $t = 1$? The velocity of the car is the amount of distance travelled (Δs) divided by the amount of time it took to travel this distance (Δt):

$$v = \frac{\Delta s}{\Delta t}$$

The velocity (rate of change) of the car at $t = 1$ cannot be calculated directly because $t = 1$ is only one instant in time, so no distance is travelled ($\Delta s = 0$), since no time has passed($\Delta t = 0$). Therefore we cannot calculate the velocity:

$$v = \frac{\Delta s}{\Delta t} = \frac{s(2) - s(2)}{2 - 2} = \frac{0}{0},$$

as division by 0 is not possible. However we can determine the average velocity between any points in time Δt. The key idea of differentiation is that we can make the difference between these points in time as small as we please, infinitesimally small.

Differentiation
In figure 4.3 we can see that the average velocity between $t = 1$ and $t = 2$ equals $\frac{\Delta s}{\Delta t}$, this is the slope of line *K*. The velocity of the car, however is the slope of the tangent line *L*. As Δt approaches zero, line *K* approaches line *L*.

Let us start by calculating the average velocity of the car between $t = 1$ and $t = 2$. The time elapsed between the two points in time can be measured by $\Delta t = 2 - 1 = 1$. The distance travelled by the car is:

FIGURE 4.3 The average velocity $\frac{\Delta s}{\Delta t}$, the slope of line K

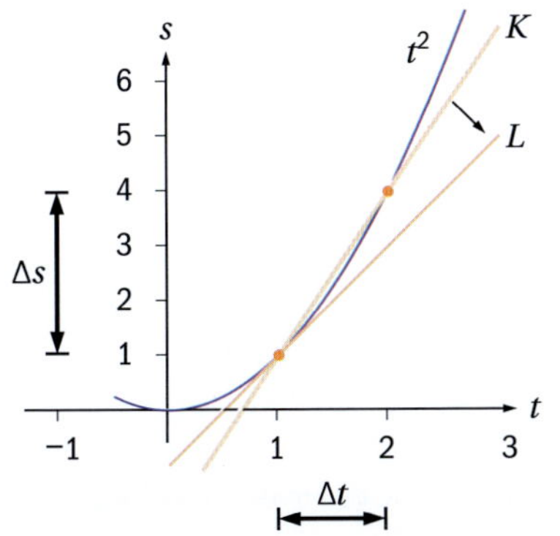

$$\begin{aligned}\Delta s &= s(1+\Delta t)-s(1)\\ &= s(2)-s(1)\\ &= 2^2-1^2\\ &= 3\end{aligned}$$

Hence the average velocity v between $t=1$ and $t=2$ equals:

$$v=\frac{\Delta s}{\Delta t}=\frac{3}{1}=3$$

Let us look at a smaller interval of time, say $\Delta t=0.1$, still starting at $t=1$. The average velocity of the car between $t=1$ and $t=1.1$ is:

$$\begin{aligned}v &= \frac{s(1+\Delta t)-s(1)}{\Delta t}\\ &= \frac{s(1.1)-s(1)}{s(1)}\\ &= \frac{1.1^2-1^2}{0.1}\\ &= \frac{0.21}{0.1}\\ &= 2.1.\end{aligned}$$

And let us go even smaller. We will make $\Delta t=0.01$. The average velocity between $t=1$ and $t=1.01$ is

$$\begin{aligned}v &= \frac{s(1.01)-s(1)}{1.01-1}\\ &= \frac{1.01^2-1^2}{0.01}\\ &= \frac{0.0201}{0.01}\\ &= 2.01.\end{aligned}$$

We can continue calculating the average velocity between smaller and smaller points in time to obtain table 4.1:

TABLE 4.1 As Δt approaches zero, v approaches 2

Δt	1	0.1	0.01	0.001	0.0001
Δs	3	0.21	0.0201	0.002001	0.00020001
$v=\frac{\Delta s}{\Delta t}$	3	2.1	2.01	2.001	2.0001

4

We see the following pattern: as Δt approaches zero, the average velocity v approaches 2. This is the velocity of the car at $t=1$. In a formula:

$$v=\frac{\Delta s}{\Delta t} \text{ as } \Delta t \text{ approaches } 0$$

For Δt approaches 0, we call $\frac{\Delta s}{\Delta t}$ the derivative of s, denoted by s'. From the previous example we see that the velocity v of the car at $t=1$ was equal to 2, hence $s'(1)=2$. This idea can be generalized to what is called the derivative of a function:

Definition 4.2 Derivative
The derivative $f'(x)$ of a function f is:

$$f'(x)=\frac{\Delta f}{\Delta x}$$

$$\frac{f(x+\Delta x)-f(x)}{\Delta x} \text{ as } \Delta x \text{ approaches } 0 \qquad (4.1)$$

Compare this definition with table 4.1: as Δt approaches zero, v approaches 2.

The derivative $f'(x)$ is the slope of the tangent line of f in x. See figure 4.4, where L is the tangent line of f in x. The slope of the tangent line is the rate of change of f in x, hence the derivative of f in x is the rate of change of f at that point x.

The slope of the curve f in x is $f'(x)$. According to Definition 4.2 this is $\frac{\Delta f}{\Delta x}$ as Δx approaches 0. In figure 4.4 you can see that $\Delta f = f(x+\Delta x)-f(x)$.

Notation of the Derivatives

The derivative f' of f can be written in various ways. The most common are the Leibniz and the Lagrange notations.

$$\underbrace{f'(x)}_{\text{Leibniz notation}} = \frac{\mathrm{d}f(x)}{\underbrace{\mathrm{d}x}_{\text{Lagrange notation}}}$$

FIGURE 4.4 The slope of the curve $f(x)$, $f'(x) = \frac{\Delta f}{\Delta x}$ as Δx approaches 0

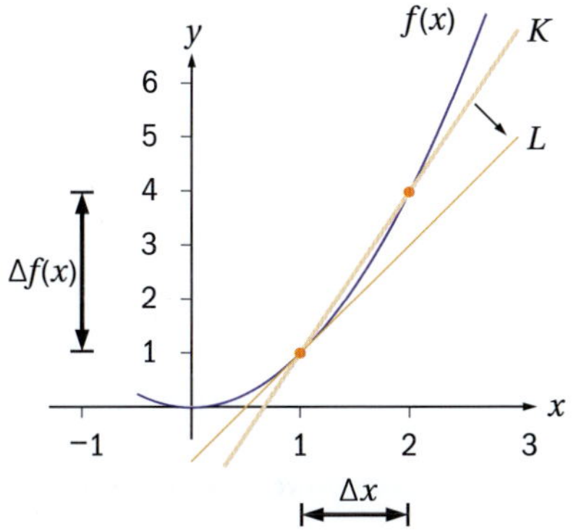

For $y=f(x)$ other notations for $f'(x)$ which are often used are:

$$\frac{dy}{dx} \text{ and } \frac{d}{dx}f(x)$$

These symbols all have the same meaning:

$$f' = f'(x) = \frac{df(x)}{dx} = \frac{d}{dx}f(x) = \frac{dy}{dx} \qquad (4.2)$$

So Definition 4.2 may be restated as

$$\frac{dy}{dx} = \frac{\Delta y}{\Delta x} \text{ as } \Delta x \text{ tends to zero} \qquad (4.3)$$

4.3 Calculating the Derivatives

Now that we understand what a derivative of a function is, we can proceed by explaining how to obtain the derivative of a function.

THEOREM 4.1

Derivative of a Constant

The derivative of any constant c is 0:

$$c' = 0 \qquad (4.4)$$

E.g., the derivative of 3, 3′, is 0.

THEOREM 4.2

Derivative of x

The derivative of x equals 1:

$$x' = 1 \tag{4.5}$$

THEOREM 4.3

Constant Multiple Rule

The derivative of a constant c multiplied by function f, $c \times f$, is the constant times the derivative of the function:

$$(c \times f)' = c \times f' \tag{4.6}$$

Example 4.2

Using the previous 2 theorems we find that the derivative of $2x$ is 2:
$(2x)' = 2 \times x' = 2 \times 1 = 2$

THEOREM 4.4

Power Rule

If $f = ax^n$, then $f' = anx^{n-1}$:

$$(ax^n)' = nax^{n-1} \tag{4.7}$$

As a mnemonic aid: bring down the power, and subtract 1 from the power.

Example 4.3

We calculate the derivative of $f = 3x^2$. We recognize $a = 3$ and $n = 2$.
Hence the derivative is:
$(3x^2)' = 2 \times 3 \times x^{2-1} = 6x$

As another example, we calculate $\sqrt{x}$. Note that $\sqrt{x} = x^{\frac{1}{2}}$. Using the power rule yields:

$$\begin{aligned}(\sqrt{x})' &= \left(x^{\frac{1}{2}}\right)' \\ &= \frac{1}{2} \times x^{-\frac{1}{2}}\end{aligned}$$

$$= \frac{1}{2} \times \frac{1}{x^{\frac{1}{2}}}$$

$$= \frac{1}{2\sqrt{x}}$$

THEOREM 4.5

Sum Rule

The derivative of a sum is the sum of the derivatives:

$$(f+g)' = f' + g' \tag{4.8}$$

4

Example 4.4

As an application of the sum rule we calculate the derivative of $5x^3 + x^2$:

$$\begin{aligned}(5x^3 + x^2)' &= (5x^3)' + (x^2)' \\ &= 3 \times 5 \times x^{3-1} + 2x^{2-1} \\ &= 15x^2 + 2x\end{aligned}$$

THEOREM 4.6

Derivative of the Exponential and Logarithm

The derivative of the exponential function e^x, where $e \approx 2.7182818\ldots$ is:

$$(e^x)' = e^x \tag{4.9}$$

And the derivative of the natural logarithm ln is:

$$(\ln x)' = \frac{1}{x} \tag{4.10}$$

THEOREM 4.7

Product Rule

The derivative of a product of functions is given by:

$$(fg)' = f'g + fg' \tag{4.11}$$

Example 4.5

As an application of the product rule we calculate the derivative of $x \ln x$:

$$(x \ln x)' = x' \ln x + x(\ln x)'$$
$$= \ln x + x \times \frac{1}{x}$$
$$= \ln x + 1$$

THEOREM 4.8

Quotient Rule

The derivative of a quotient of functions is given by:

$$\left(\frac{f}{g}\right)' = \frac{f'g - fg'}{g^2} \quad (4.12)$$

4

Example 4.6

As an application of the quotient rule we calculate the derivative of $\frac{\ln x}{x}$:

$$\left(\frac{\ln x}{x}\right)' = \frac{(\ln x)' x - \ln x \times x'}{x^2}$$
$$= \frac{\frac{1}{x} \times x - \ln x \times 1}{x^2}$$
$$= \frac{1 - \ln x}{x^2}$$

THEOREM 4.9

Chain Rule

The derivative of a chain of functions f and g, $f(g(x))$ is given by:

$$(f(g(x)))' = f'(g(x))g'(x) \quad (4.13)$$

Or, in Lagrange notation:

$$\frac{\mathrm{d}f(g(x))}{\mathrm{d}x} = \frac{\mathrm{d}f}{\mathrm{d}g}\frac{\mathrm{d}g}{\mathrm{d}x} \quad (4.14)$$

Example 4.7

As an application of the chain rule, we calculate the derivative of $\ln x^3$:
Set $g(x) = x^3$ and $f(g) = \ln(g)$. Then:

$$(f(g(x)))' = g'(x)f'(g(x))$$

$$(\ln x^3)' = (x^3)' \times \ln'(x^3)$$

$$= 3x^2 \times \frac{1}{x^3}$$

$$= \frac{3}{x}$$

You might prefer to apply the chain rule using the Lagrange notation, as in formula (4.14):

$$\frac{\mathrm{d}f(g(x))}{\mathrm{d}x} = \frac{\mathrm{d}f}{\mathrm{d}g}\frac{\mathrm{d}g}{\mathrm{d}x}$$

$$\frac{\mathrm{d}\ln(x^3)}{\mathrm{d}x} = \frac{\mathrm{d}\ln(g)}{\mathrm{d}g} \times \frac{\mathrm{d}x^3}{\mathrm{d}x}$$

$$= \frac{1}{x^3} \times 3x^2$$

$$= \frac{3}{x}$$

4

4.4 Minima and Maxima

We are often interested in the minimum and the maximum of a function. For example, you would like to buy something when the price is at its minimum, and sell something when it is at its maximum. The next theorem will be followed by an example.

FIGURE 4.5 The function f attains its minimum or maximum when the target line is horizontal, i.e, when the derivative f' is zero

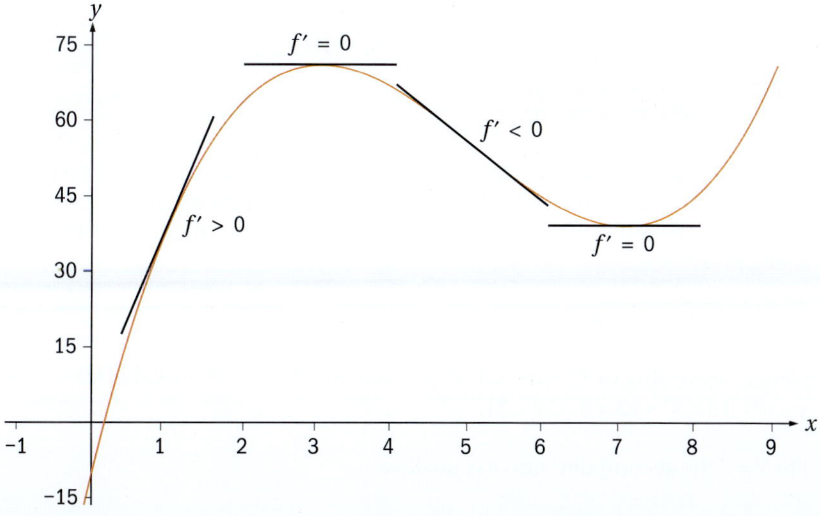

In figure 4.5 we can see that where the function f attains its minimum or maximum, the tangent line is horizontal. The slope of this tangent line equals zero. This means the derivative of f is zero if f is at its maximum or minimum.

THEOREM 4.10

Maxima and Minima

A function $f(x)$ attains its maximum in $x = x_0$ if
1 $f'(x_0) = 0$
2 $f''(x_0) < 0$

And a function $f(x)$ attains its minimum in $x = x_0$ if
1 $f'(x_0) = 0$
2 $f''(x_0) > 0$

Explanation
We will explain why 1 and 2 imply that f attains its maximum. Since $f''(x_0) < 0$ the derivative $f'(x)$ is decreasing around x_0, meaning the speed at which f grows is decreasing. $f'(x_0) = 0$ means f has speed 0 in x_0. Both arguments combined indicate that the speed of f first decreases, then becomes zero in x_0, and continues decreasing. This can only happen when f attains a maximum in x_0. The explanation of the minimum is similar.

A function is increasing when its derivative is positive, because the rate of change, and hence the slope is positive. A function is decreasing when its derivative is negative. When a function is at its maximum or minimum its rate of change is neither positive nor negative, it is zero. Hence we can find the maxima and minima of functions where the derivative is zero.

In figure 4.6 we can see there is a local maximum in the vicinity of $x = 3$, and a local minimum in the vicinity of $x = 7$. Both of these points can be found by setting the derivative of $y = f(x)$ equal to zero:

$$\begin{aligned} f'(x) &= 0 \\ (x^3 - 15x^2 + 63x - 10)' &= 0 \\ 3x^2 - 30x + 63 &= 0 \end{aligned}$$

This quadratic equation has solutions $x = 3$ and $x = 7$. So y attains a maximum or a minimum in $x = 3$ and in $x = 7$. But how can we discern the maximum from the minimum? For $x = 3$ the second derivative of f, $f''(x) = 6x - 30$, is negative:

$$6 \times 3 - 30 = -12 < 0$$

Hence, according to Theorem 4.10, y is maximal for $x_0 = 3$. Its value is: $y = 3^3 - 15 \times 3^2 + 63 \times 3 - 10 = 71$.

For $x = 7$ the second derivative is positive:

$$6 \times 7 - 30 = 12 > 0$$

FIGURE 4.6 $y = x^3 - 15x^2 + 63x - 10$

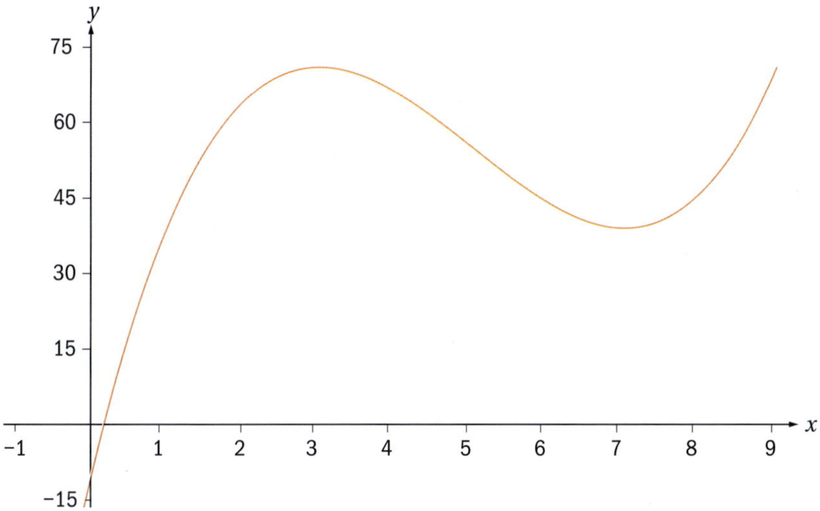

Hence, according to Theorem 4.10, y is minimal for $x_0 = 7$. Its value is:
$y = 7^3 - 15 \times 7^2 + 63 \times 7 - 10 = 39$.

4.5 Price Elasticity of Demand

Marketers must understand the relationship between price and demand for a product. This is crucial, especially when a firm is introducing new products or services to the market, or when a firm is changing its pricing. The price elasticity of demand is a tool for understanding this relationship. There are many methods of calculating the price elasticity of demand, however, in this book we will only be using the arc method.

The Apple iPad and the Price Elasticity Equation

Steve Jobs (1955–2011)

An iconic brand, borne from the image of a fruit. Who said money can't grow on trees like apples do? For Steve Jobs, the CEO of Apple Inc., building the largest consumer base in the industry and offering the most innovative products in the market were the main targets, which he had evidently been achieving since the phenomenal launch of the iPod in 2001.

The iPad was one of a new generation of tablet computers that are set to replace laptops, books, and revolutionize the technological industry. It was released in April 2010, and sold more than 3 million units in 80 days. This great sales turnover wasn't a surprise to many, as Apple had a clear strategy that had been in operation years before the launch of this product. Apple had announced years before that they were going to release the iPad, in order to stir up hype and increase demand for the product, which they realized with great success. However, Apple, never disclosed any information about the price range the product was going to be in and speculations about the price of the iPad increased in the market. Due to the very high specifications of the product and the features it was meant to include, it was estimated that it would be sold at $900 plus. Since Apple managed to sell a vast number of units on the first day of the launch, the price elasticity of the product is worth thinking about. Apple had released iPads that were priced way below the range of speculated market prices, and consumers were flocking to stores and ordering in mass online. One question however: were those people just typical Apple devotees, who buy everything Apple launches regardless of the price?

Source: www.termpaperwarehouse.com

Definition 4.3 Price Elasticity of Demand
The price elasticity of demand is a measure of the sensitivity of the quantity demanded of goods to change in price.

As an example, we will consider the quantity demanded of a product at two different prices. If the price is set to € 1, the company can sell 7 units this week; when the price is set to € 3, the company sells 5 units, see table 4.2.

TABLE 4.2

P	Q_d
1	7
3	5

The relative increase in price is thus given by:

$$\frac{\Delta p}{\text{mid}\,P} \times 100\%$$

Where ΔP is the change in price, and mid P the midpoint of the two prices considered. In this example $\Delta P = 3 - 1 = 2$, and mid$P = \frac{1+3}{2} = 2$. Hence

$$\frac{\Delta P}{\text{mid}\,P} \times 100\% = \frac{2}{2} \times 100\% = 100\%$$

The relative increase in demand is given by:

$$\frac{\Delta Q_d}{\text{mid}\,Q_d} \times 100\%$$

Where ΔQ_d is the change in quantity demanded, and mid Q_d is the midpoint of the two quantities. In this example $\Delta Q_d = 5 - 7 = -2$, and mid $Q_d = \frac{7+5}{2} = 6$. Hence

$$\frac{\Delta Q_d}{\text{mid } Q_d} \times 100\% = \frac{-2}{6} \times 100\% = -33.333\%$$

Now, if the price increases from 1 to 3 euros, which is an increase of 100% relative to the midpoint of 1 and 3 euros, then the demand decreases by 2 units, which is a decrease of 33.333% relative to the midpoint of 5 and 7 pieces. The absolute value of the ratio of the percentage changes for quantity and price gives us an indication of how much demand is influenced by price. This absolute value of the ratio is called the price elasticity of demand. In this example:

$$PED = \left| \frac{\frac{\Delta Q_d}{\text{mid } Q_d} \times 100\%}{\frac{\Delta P}{\text{mid } P} \times 100\%} \right| = \left| \frac{-33.333\%}{100\%} \right| = 0.333 \qquad (4.15)$$

4

In general we have:

Definition 4.4 Price Elasticity of Demand (Arc Method)
The price elasticity of demand (*PED*) is the absolute value of the ratio of percentage of change in the quantity demanded and the percentage of the change in price. In a formula:

$$PED = \left| \frac{\frac{\Delta Q_d}{\text{mid } Q_d} \times 100\%}{\frac{\Delta P}{\text{mid } P} \times 100\%} \right|$$

See Formula (1.9) in Theorem 1.2 for the definition of the absolute value sign $|\cdot|$.

Using one of the properties of fractions (see Formula 1.3 in Theorem 1.1), this formula can be simplified as:

$$PED = \left| \frac{\Delta Q_d}{\Delta P} \times \frac{\text{mid } P}{\text{mid } Q} \right| \qquad (4.16)$$

Table 4.2 gives us $\Delta Q_d = 5 - 7 = -2$, mid $Q_d = \frac{7+5}{2} = 6$, $\Delta P = 3 - 1 = 2$, and mid $P = \frac{1+3}{2} = 2$. Inserting these figures into formula (4.16) yields:

$$PED = \left| \frac{\Delta Q_d}{\Delta P} \times \frac{\text{mid} P}{\text{mid} Q} \right| = \left| \frac{-2}{2} \times \frac{2}{6} \right| = 0.333 \qquad (4.17)$$

This is the same result as before, in formula (4.15).

If *PED* > 1, then the price is elastic, meaning that a slight change in price affects the demand considerably. An example of a product with an elastic price is milk. If milk were to become too expensive, you would probably buy something else.

If *PED* < 1, then the price is inelastic, meaning that the change in price has little influence on demand. An example of a product with an inelastic price is electricity: if electricity were to become more expensive, you would probably still use it, as you can't get your iPad to run on another source of energy.

If *PED* = 1 then the price is unit elastic, meaning that the increase in price will result in the same decrease in the percentage of demand.

Summary

- $f'(x) = \dfrac{f(x+\Delta x) - f(x)}{\Delta x}$ as Δx approaches 0.

- Rules for differentiation:

  - $c' = 0$
  - $(\ln x)' = \dfrac{1}{x}$
  - $(e^x)' = e^x$
  - Constant multiple rule $(c \times f)' = c \times f'$
  - Power rule $(ax^n)' = nax^{n-1}$
  - Sum rule $(f+g)' = f' + g'$
  - Product rule $(fg)' = f'g + fg'$
  - Quotient rule $\left(\dfrac{f}{g}\right)' = \dfrac{f'g - fg'}{g^2}$
  - Chain rule $(f(g(x)))' = g'(x)f'(g(x))$

    or in Lagrange notation:

    $\dfrac{\mathrm{d}f(g(x))}{\mathrm{d}x} = \dfrac{\mathrm{d}f}{\mathrm{d}g}\dfrac{\mathrm{d}g}{\mathrm{d}x}$

- A function $f(x)$ attains its maximum in $x = x_0$ if:

  1 $f'(x_0) = 0$
  2 $f''(x_0) < 0$

- A function $f(x)$ attains its minimum in $x = x_0$ if:

  1 $f'(x_0) = 0$
  2 $f''(x_0) > 0$

- The price elasticity of demand is given by:

$$PED = \left| \frac{\dfrac{\Delta Q_d}{\text{mid } Q_d} \times 100\%}{\dfrac{\Delta P}{\text{mid } P} \times 100\%} \right|$$

4

Exercises

4

4.1 Differentiate the following functions, using the power rule:

a 3
b $3x^1$
c $3x$
d $2x$
e $5x^3$
f $3x^2$
g $2x^3$
h $3x^4$
i $x^{0.5}$
j $\sqrt{x}$
k x^{-1}
l $\frac{1}{x}$
m x^{-2}
n $\frac{1}{x^2}$
o $\frac{5}{x^4}$
p $5x^{-4}$

4.2 Using the sum rule, find $f'(x)$, when $f(x)$ is given by:

a $3x-2$
b $3x^2-4$
c e^x+2x
d $2x+x^2$
e x^3+x^2+x+1
f $x+\ln x$
g $2x^3+5\ln x-1$
h $3x^2+2x^3$
i $\ln x-e^x-1$

4.3 Find the derivatives of the following functions, using the product rule:

a $x\times x$
b $x\ln x$
c $x^2\ln x$
d $(x-1)(x+1)$
e $\sqrt{x}\ln x$
f $(\sqrt{x}-123)(\sqrt{x}+123)$
g $\ln x\times\ln x$
h $\sqrt{x}\times\sqrt{x}$
i $2x(\ln x+\sqrt{x})$

4.4 Find the derivatives of the following functions, using the quotient rule:

a $\frac{x}{x+1}$

b $\frac{x+1}{x+2}$

c* $\frac{x+1}{x^2-1}$

d $\frac{\ln x}{x}$

e $\frac{\ln x}{\sqrt{x}}$

f* $\frac{\sqrt{x}}{x+1}$

4

4.5 Differentiate the following functions, using the chain rule:

a $(x+1)^3$
b $\sqrt{x+1}$
c $\sqrt{2x^3}$
d $\ln \sqrt{x}$
e $\ln x^{0.5}$
f e^{-x^2}
g $\ln x^2+x+1$
h $\frac{1}{1-x^2}$
i e^3x

4.6 Find the maxima or minima of:

a $2x+x^2$
b $x-\ln x$
c $x\times x$
d $(x-1)(x+1)$
e $\ln x\times \ln x$
f $\frac{\sqrt{x}}{x+1}$
g $\ln x^2+x+1$
h e^x+e^{-x}
i e^{-x^2}

4.7* Find the maximum or minimum of the parabola $y=ax^2+bx+c$. Compare your result with Exercise 3.6 in chapter 3.

4.8* Derive the sum rule using the definition of the derivative.

4.9* **a** Show that $a^x=e^{x\ln a}$.
b Use the previous result to find the derivative of a^x.

5 Economic Application of Functions and Differentiation

At the end of this chapter you will be able to:
- understand the difference between marginal cost and opportunity cost
- calculate marginal costs, marginal profit and marginal revenue
- analyse marginal cost, marginal profit and marginal revenue
- calculate the break-even point

This chapter mostly discusses the application of derivatives and differentiations. The key concept is to finding the point where functions take on their maximum or minimum value, and also therefore the derivatives that are used to find those values. More specifically, we will be dealing with marginal cost, marginal profit and marginal revenue.

Application of Differentiation

Economic research often uses calculus to examine functional relationships. An example of this includes the relationship between the dependent variable income and various predictors, or independent variables, such as education and experience. If average income rises as years of education and work experience increase, then a positive relationship exists between the variables, namely that income is a function of education and experience. Differential calculus, the process of obtaining derivatives, enables economists to measure the average change in income relative to a single year's increase in education and/or experience.

According to Harvard economist Greg Mankiw, author of *Principles of Economics*, Mankiw writes that economists use the term 'marginal changes' to describe small, incremental changes, such as changes in work hours or factory output.
Determining marginal revenues and costs can help business managers maximize their profits and measure the rate of increase in profit that results from each increase in production. As long as marginal revenue exceeds marginal cost, the firm increases its profits.

Source: www.ehow.com

5.1 Cost Concepts

The mother of all economic ideas is the cost-benefit principle. It says you take action if – and only if – the extra benefit from taking it is greater that the extra cost. Could a principle be any more simple? Still, it is not always easy to apply. For example you are about to buy a €12 calculator at the shop at the campus next door when a friend tells you that the same calculator is available for €6 at a media market in the town. Do you go town and get the calculator for €6? Or do you buy it at the shop at the campus? Note that in both cases, if the calculator functions under warranty, you can send it back to the manufacturer for repair (Franck, 2007).

As life involves a series of decisions, consumers and businesses face questions such as whether to put in a few extra hours at work, save a little extra each month, buy a new computer, or build an additional production facility (source: www.ehow.com).

In this section we will discuss a few cost concepts. The major cost concepts are: Fixed costs (*FC*), Variable costs (*VC*), Total costs (*TC*), Marginal costs (*MC*) and Opportunity costs (*OP*) (see figure 5.1).

5

FIGURE 5.1 The major cost concepts

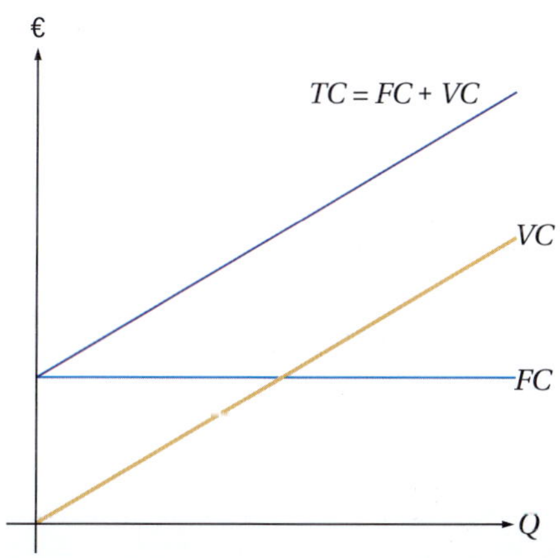

5.2 Revenue, Cost, Profit and Break-Even Point

Imagine you run a small store selling books. The total cost per month for running this store is equal to the rent, plus the number of books purchased multiplied by the price of those books. The rent is a fixed cost: it needs to be paid every month no matter how many books are bought or sold. The cost of books is variable; the more books the store purchases, the more the manager has to pay. If the rent is €1,000 and the unit selling price (*USP*) is €50 per book, and if this store has purchased 200 books this month, then the total cost *TC* for this month is equal to:

$$\begin{aligned} TC &= 1{,}000 + 200 \times 50 \\ &= 11{,}000 \end{aligned}$$

And in general:

THEOREM 5.1

Total Costs

The total cost TC is given by

$$TC = FC + VC \qquad (5.1)$$

Where VC is the variable cost. Note that VC equals $USP \times Q$, with USP the unit selling price, and Q the quantity of items bought.

If this bookstore sold these $Q = 200$ books this month for $USP = €60$ each, then the amount of cash in the cash register at the end of this month, TC, equals €60 times 200 = €12,000. This is the price of a book multiplied by the number of books sold.

5

THEOREM 5.2

Total Revenue

The total revenue TR equals the price P times the quantity Q sold:

$$TR = PQ \qquad (5.2)$$

The total profit TP this bookstore made this month equals their revenue minus their total costs: € 12,000 − € 11,000 = € 1,000.

THEOREM 5.3

Total Profit

The total profit TP equals the total revenue TR minus the total cost TC:

$$TP = TR - TC \qquad (5.3)$$

If this bookstore bought and sold only 70 books this month, then their total revenue would be less than their total costs:

$$\begin{aligned} TC &= FC + VC \\ &= 1{,}000 + 70 \times 50 \\ &= 4{,}500 \end{aligned}$$

Whereas their total revenue equals:

$$\begin{aligned} TR &= PQ \\ &= 60 \times 70 \\ &= 4.200 \end{aligned}$$

Hence their total profit is negative:

$$\begin{aligned} TP &= TR + TC \\ &= 4.200 - 4.500 \\ &= -300 \end{aligned}$$

which is a loss of €300. In the example preceding Theorem 5.3, the total profit was €1,000 because the store sold 200 books. Hence they need to sell somewhere between 70 and 200 books to achieve zero profit. This amount of books sold is called the break-even point (see figure 5.2).

Definition 5.1 Break Even
The break-even point Q_b is a quantity whereby the total profit is zero. That is when the total costs equal the total revenue. Hence the break-even point Q_b is the solution of either:

$$TP = 0 \tag{5.4}$$

or

$$TR = TC \tag{5.5}$$

FIGURE 5.2 Breakevenpoint: *TC* equals *TR*, total profit is zero

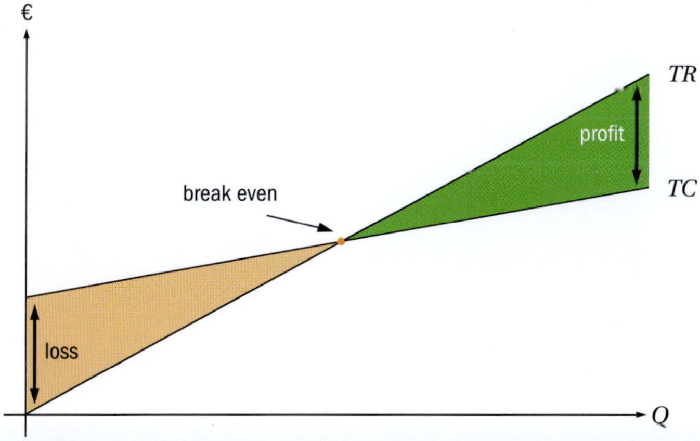

Example 5.1

The break-even point Q_b for the bookstore is the solution of
$60Q_b = 1{,}000 + 50Q_b$

This solution is $Q_b = 100$. They must sell 100 books just to cover the costs. Selling more than 100 books a month would generate a profit, but fewer would generate a loss.

5.3 Maximizing Profit and Minimizing Cost

In this section we will use the following example to explain how to maximize the total profit TP. We will see what the marginal revenue and the marginal cost are, and how to use them to maximize the profit (see figure 5.3).

Example 5.2

Suppose the price P of a product is given by
$P(Q) = Q^2 - 15Q + 83$

Thus the price P depends on the quantity Q. Furthermore, let the total cost function $TC(Q) = FC + VC$ be given by:

$$TC(Q) = 10 + 20Q \tag{5.6}$$

Thus the fixed cost is 10, and the variable cost is 20. We are interested in finding the quantity Q_0 such that the profit TP is maximized. Since $TP = TR - TC$ we must find TR first. The revenue TR equals:

$$\begin{aligned} TR(Q) &= P(Q) \times Q \\ &= (Q^2 - 15Q + 83)Q \\ &= Q^3 - 15Q^2 + 83Q \end{aligned} \tag{5.7}$$

Now TP equals:

$$\begin{aligned} TP(Q) &= TR(Q) - TC(Q) \\ &= Q^3 - 15Q^2 + 83Q - 10 - 20Q \\ &= Q^3 - 15Q^2 + 63Q - 10 \end{aligned}$$

FIGURE 5.3 The profit function TP

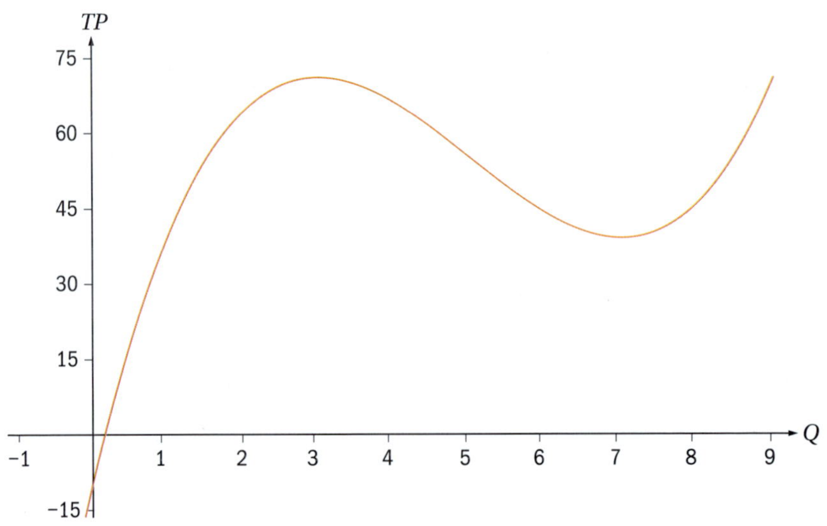

The rest of this example is the same as section 4.4 on minima and maxima. In figure 5.3 we can see, there is a maximum in the vicinity of $Q=3$, and a minimum in the vicinity of $Q=7$. Both of these points can be found by setting the derivative of TP equal to zero:

$$TP'=0$$
$$(Q^3-15Q^2+63Q-10)'=0$$
$$3Q^2-30Q+63=0$$

This quadratic equation has solutions $Q=3$ and $Q=7$, hence the profit attains a maximum or a minimum in $Q=3$ and $Q=7$. In order to check whether TP is maximal or minimal in $Q=3$ and $Q=7$ we have to calculate the second derivative of TP:

$$\begin{aligned} TP'' &= (TP')' \\ &= (3Q^2-30Q+63)' \\ &= 6Q-30 \end{aligned}$$

For $Q=3$ the second derivative is negative:
$6\times 3-30=-12<0$

Hence, according to Theorem 4.10, the profit is maximal. For $Q=3$ the profit TP equals $3^3-15\times 3^2+63\times 3-10=71$.
For $Q=7$ the second derivative is positive:
$6\times 7-30=12>0$

Hence, according to Theorem 4.10, the profit is minimal for $Q=7$. For $Q=7$ the profit TP equals $7^3-15\times 7^2+63\times 7-10=39$.

This whole procedure can be generalized using the following theorems:

THEOREM 5.4

Maximal Profit

The profit function $TP(Q)$ is maximized for Q_0 such that

1 $TP'(Q_0)=0$
2 $TP''(Q_0)<0$

THEOREM 5.5

Minimal Cost

The cost function $TC'(Q_0)=0$ is minimized for Q_0 such that

1 $TC'(Q_0)=0$
2 $TC''(Q_0)>0$

5

5.4 Marginal Revenue and Cost

The market structure is often divided in four types: Monopoly, Oligopoly, Perfect competition and Monopolistic competition. The relation between marginal revenue and the quantity of output produced depends on the market structure.

Definition 5.2 Marginal Cost and Marginal Revenue
The *marginal* cost function $MC(Q)$ is defined as the derivative of the cost function $TC(Q)$:

$$MC(Q) = TC'(Q) \tag{5.8}$$

Marginal revenue is the extra revenue made when a firm sells one extra unit of output. The marginal revenue function $MR(Q)$is defined as the derivative of the revenue function $TR(Q)$:

$$MR(Q) = TR'(Q) \tag{5.9}$$

5

Figure 5.4 shows the curve of the total revenue, the curve of the marginal cost and the total profit. Profit is maximized when the slope of the total revenue curve (marginal revenue) equals the slope of the total cost curve (marginal cost).

FIGURE 5.4 Curves showing total revenue, total cost and total profits

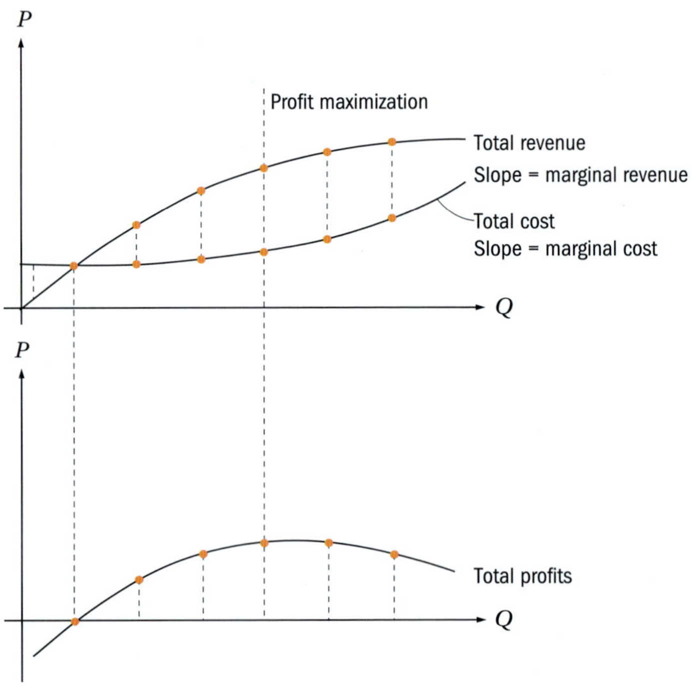

Source: www.sparkcharts.sparknotes.com

These definitions can be used in the following theorem:

THEOREM 5.6

Maximal Profit and Minimal Costs

The profit is maximized when:

1 $MR = MC$
2 $MR' < MC'$

And the costs are minimized when:

3 $MC = 0$
4 $MC' > 0$

Explanation: using Theorem 5.4, part 1 follows from $TP' = MR - MC = 0$, part 2 from $TP'' = MR' - MC' < 0$. Furthermore, parts 3 and 4 follow directly from Theorem 4.10.

5

Example 5.3

In Example 5.1 the revenue TR was given by formula 5.7:
$TR(Q) = Q^3 - 15Q^2 + 83Q$

and the cost TC by formula 5.6:
$TC(Q) = 10 + 20Q$

In order to apply Theorem 5.6, we need to calculate the marginal revenue, the marginal cost, and their derivatives:
$MR = TR' = 3Q^2 - 30Q + 83$
$MC = TC' = 20$
$MR' = (3Q^2 - 30Q + 83)' = 6Q - 30$
$MC' = 20' = 0$

According to Theorem 5.6 the profit is maximized when:
$$3Q^2 - 30Q + 83 = 20 \qquad (5.10)$$

and
$$6Q - 30 < 0, \text{ thus } Q < 5 \qquad (5.11)$$

The solutions of equation (5.10) are $Q = 3$ and $Q = 7$. Since only $Q = 3$ satisfies condition (5.11) the profit is maximized for $Q = 3$.

5.5 Opportunity Cost

Opportunity cost is a critical concept in economics, business and management. It serves as a crucial guide to making the best decisions possible. The main goal of business is to maximize profit and to minimize opportunity cost. So what does opportunity cost actually mean?

The opportunity cost refers to the benefit or cost of something that must be given up in order to get something else. Its benefit could be given in terms of money, time, satisfaction, etc.

Why is the opportunity cost important?
- Choice involves sacrifice. The more clothes you choose to buy, the less money you will have to spend on other goods.
- The more wine France produces, the fewer resources France will have for producing other goods. In other words the production or consumption of one type of goods involves the sacrifice of alternatives.

Example 5.4

By taking a plane, Sylvia can travel from Amsterdam to Paris in one hour. The same trip takes Sylvia 5 hours by bus. The airfare is €90 and the bus fare is €30. When Sylvia is not travelling she can work and earn €30 per hour.

a What is the opportunity cost (OC) for Sylvia for traveling by bus?
b What is the OC for Sylvia for traveling by plane?
c Which is the cheaper mode of travel for Sylvia?

Source: http://myweb.lmu.edu

Solution:

a
$$\begin{aligned} OC_{\text{bus}} &= \text{Direct Cost} + \text{Indirect Cost} = \text{Bus Fare} + \text{Time Cost} \\ &= €30 + (5\ \text{hours})(€30/\text{hour}) \\ &= €30 + €150 = €180 \end{aligned}$$

b $OC_{\text{plane}} = €90 + (1\ \text{hours})(€30) = €120$

c Taking the plane would be cheaper for Sylvia.

5

Summary

- Total cost is the fixed cost plus the variable cost:

  $TC = FC + VC$

- Total revenue is price times quantity:

  $TR = PQ$

- Total profit equals total revenue minus total cost:

  $TP = TR - TC$

- The break-even point occurs where:

  $TP = 0$ or $TR = TC$

- Marginal cost MC equals:

  $MC = TC'$

- Marginal revenue MR equals:

  $MR = TR'$

- The profit is maximized when:

  1 $TP' = 0$
  2 $TP'' > 0$

  or when

  1 $MP = 0$
  2 $MP' < 0$

- And the costs are minimized when:

  1 $TC' = 0$
  2 $TC'' > 0$

  or when

  1 $MR = MC$
  2 $MR' < MC'$

- The opportunity cost refers to the benefit or cost of something that must be given up in order to get something else.

5

Exercises

5.1 Assume a fixed cost of €900, a variable cost of €4.50 and a selling price of €5.50.

a What is the break-even point?
b How many units must be sold to make a profit of €500?
c How many units must be sold to have an average of €0.25 profit per unit? €0.50 per unit? €0.75 per unit?

5.2 A firm is selling two products – chairs and bar stools – each at $50 per unit. Chairs have a variable cost of $25 and bar stools have a variable cost of $20. Fixed costs for the firm are $20,000.

a If the sales mix is 1:1 (one chair sold for every bar stool sold), what is the break-even point in units of chairs and bar stools? And in dollars of sales?
b If the sales mix changes to 1:4 (one chair sold for every four bar stools), what is the break-even point in units of chairs and bar stools? And in dollars of sales?

5.3 Consider a product for which the fixed costs for producing it are €68. The unit cost is €3.00 for each batch of 200 units. Find the quantity that would correspond to a cost of €200.

5.4 An apartment complex has 250 apartments for rent. If they rent Q apartments then their monthly profit is given by:

$$TP(Q) = -8Q^2 + 3{,}200Q - 80{,}000$$

How many apartments should they rent in order to maximize their profit?

5.5 A production facility is capable of producing 60,000 widgets in a day and the total daily cost TC of producing Q widgets in a day is given by:

$$TC(Q) = 250{,}000 + 0.08Q + \frac{200{,}000{,}000}{Q}$$

How many widgets per day should they produce in order to minimize production costs?

5.6 The production costs C per week for producing Q widgets are given by:

$$C = 500 + 350Q - 0.09Q^2, \qquad 0 < Q \leq 1{,}000$$

a Determine the cost of producing the 301st widget, when the 300th is already produced.
b What is the rate of change of the cost at $Q = 300$?

c* Use the relation $\frac{\Delta C}{\Delta Q} \approx \frac{dC}{dQ}$ to approximate **a** from **b.**

5.7 A pizza firm determines that at a price of €1.00 per slice, they'll sell an average of 400 slices per night. If they increase the price to €1.40, the average number of slices sold per night would be 200.

a Determine the demand relation, assuming it is linear.
b Find the price that maximizes revenue.
c Find the quantity that maximizes revenue.
d Find how much the total revenue would change by if the price was increased from €0.80 to €0.90 per slice.

5.8 A firm produces and sells Q_q items of a certain product per day. Production costs are given by $TC(Q) = 200 + Q^{1.5}$ in kiloeuros (k€, 1,000 euros) per day.

The market price P at which the product is sold is 50 k€ per item.

a Give an expression for the profits $TR(Q)$ on this product.
b Calculate $TP(20)$.
c Calculate the marginal cost TC' for $Q = 36$.

5

5.9 A car wholesaler sells 100,000 cars per year and reorders them in batches of x. It costs €1,000 to place each order, and €200 per year in storage charges for each car. Calculate the optimum size of each order in order to minimize the total costs (order + storage costs).

5.10* A company increases the prices of its products by 5%.

a By how much can it accommodate a decrease in the quantity sold in order to have the same revenue?
b Assume that the variable cost is €7, the old price is €5, and fixed cost is €20,000. By what percentage does the quantity have to change to achieve the same profit?

5.11 Calculate the price elasticity of the following prices and demands.

a If an item has a price of €30 the demand is 80,000 pieces; if this item has a price of €40 the demand is 60,000 pieces.
b If a Ferrari costs €180,000, 6 of them would be sold, if the price were to drop to €100,000 the demand would be 8 units.

5.12 A country is capable of producing the following combinations of goods and services per period of time, assuming that it makes full use of its resources of land, labour and capital:

Goods (units)	100	80	60	40	20	0
Services (units)	0	50	90	120	140	150

a Draw the production possibility curve for this country.
b What is the opportunity cost (in terms of services) of producing 20 extra units of goods when this country is initially producing:
i 60 units of goods
ii 20 units of goods

5.13 Assume that country A devotes all of its resources to a particular activity: it can either produce 80 units of wheat or 80 units of barley. If country B does the same, it can either produce 30 units of wheat or 60 units of barley. What is the opportunity cost for country B if it decides to produce 20 units of barley?

5.14 An exercise on marginal functions. The following table shows the total number of visits that Eleanor makes to the cinema per week.

Visits	**1**	**2**	**3**	**4**	**5**	**6**	**7**	**8**
TU (€)	12	20	25	28	30	31	31	29
MU(€)								

a Copy the table and fill in the figures for marginal utility.
b Draw a graph of the figures for total and marginal utility.
c How many visits to the cinema will she make per week if the price of a ticket is:
i €4.00
ii €2.50

5.15 **a** Copy the table and fill in the costs for a firm. (Note: enter the figures in the *MC* column *between* outputs of 0 and 1, 1 and 2, 2 and 3, etc.)

Output	***TC* (€)**	***AC* (€)**	***MC* (€)**
0	55		
1	85		
2	110		
3	130		
4		40	
5		42	
6	280		
7			90
8			110
9	610		
10			150

b How much is the total fixed cost at:
  i an output of 0?
  ii an output of 6?
c How much is the average fixed cost at:
  i an output of 5?
  ii an output of 10?
d How much is the total variable cost at an output of 5?
e How much is the average variable cost at an output of 10?

6

Σummation, Percentages and Interest

At the end of this chapter you should be able to:
- understand the concept of sigma sum
- understand the concept of percentage
- understand the difference between an increase or a decrease of percentage
- manipulate the scale factor of percentages
- solve simple and compound interest problems
- understand the concept of depreciation
- understand the concept of time value of an investment
- understand the concept of value of investments

In today's business, it is important to understand and to be able to manipulate percentage and interest because they are parts of our daily routine. Take for example the summer sale, where nearly all shops discount their products. Nowadays the concept of borrowing and lending is a large part of all nations' economies. You often see ads saying: 'buy now and pay later'. Whether you are an individual, a firm or a government, loans and interest will play a big role in your business.
This chapter will explore the main concepts of percentages, loan and interests.

iPad retains value longer than Kindle (CNN) – Technology is not like wine

Electronics almost never gain value after they've been sitting around for a while, but some stand the test of time better than others. As Apple is reportedly readying a third-generation iPad to be unveiled in the next month or so, owners of previous versions of the tablet may be contemplating a way to unload them for cash. Fortunate for them, the iPad can be resold for about half its original price a year or more after it hit the market, according to data from two popular online researchers that was compiled for CNN.
On the other hand, Amazon.com's Kindle devices, which analysts say pose the greatest threat to Apple's dominance in tablets, have not been as sought after in the secondary market.
The various Kindle e-readers generally are worth between 25% and 33% of their original price a year after their release, according to data from electronics reseller Gazelle.
The iPad costs more than the Kindle, but the Apple tablet retains more of its value over time, according to data.

After a year, the Kindle's value can sink even faster. The Kindle 3, a $2\frac{1}{2}$-year-old product still being sold by Amazon for €139 under the name Kindle Keyboard, can be worth far less. Amazon didn't respond to a request for comment on these statistics. Likewise, an Apple spokeswoman declined to comment. One might assume that because the iPad is more expensive than even the top-of-the-line Kindle (€499 to the Kindle Fire's €199) that Apple's tablet is likely to depreciate more quickly. But Kindles, on average, depreciate 22% faster than iPads, Gazelle executive Anthony Scarsella wrote in an e-mail.

6.1 Σummation Sign Σ

There are many situations in which we need to sum up a large amount of numbers. Because this happens so often, mathematicians have adopted a notation using the Greek letter Σ (the Greek capital S, for Sum).

Definition 6.1 Summation Sign Σ

$$\sum_{i=1}^{n} x_i$$

This means: sum of all x_i-values where index i goes from 1 to n, i.e.:

$$\sum_{i=1}^{n} x_i = x_1 + x_2 + \ldots + x_n \qquad (6.1)$$

Let us show a summation using definition 6.1 step by step:

1 Take all the numbers between 1 and n, i.e.

$$i = 1, 2, \ldots, n$$

2 Write

$$x_1, x_2, \ldots, x_n$$

thus replacing the i with all the numbers between 1 and n.

3 Because Σ means summation, now write

$$x_1 + x_2 + \ldots + x_n$$

thereby replacing the commas with plus signs.

Example 6.1

The summation

$$\sum_{i=1}^{3} 2i$$

can be found by

1 taking all the numbers between 1 and 3:
1, 2, and 3
2 Writing
$2 \times 1, 2 \times 2, 2 \times 3$
thereby replacing the i with all the numbers between 1 and 3.
3 Because Σ means summation, we now write:
$2 \times 1 + 2 \times 2 + 2 \times 3$,
thereby replacing the commas with plus signs. The result is:
$2 \times 1 + 2 \times 2 + 2 \times 3 = 2 + 4 + 6 = 12$

Example 6.2

$$\sum_{i=3}^{3} x_i = x_1 + x_2 + x_3$$

$$\sum_{i=1}^{3} i = 1 + 2 + 3 = 6$$

$$\sum_{k=10}^{11} k = 10 + 11 = 21$$

$$\sum_{k=1}^{4} k^2 = 1^2 + 2^2 + 3^2 + 4^2 = 30$$

$$\sum_{i=0}^{3} 2^i = 2^0 + 2^1 + 2^2 + 2^3 = 15$$

$$\sum_{i=1}^{3} (2 + 3i) = (2 + 3) + (2 + 6) + (2 + 9) = 24$$

6.2 Percentages

6

The concept of percentages is very important in the business world. Most of the time managers give their outcome in terms of percentages. For example, a 50% decrease in sales, a 75% increase in profit, a 3% increase in market share. Also growth rates, such as interest rates, are often expressed in percentages.
'Per cent' is derived from the Latin 'per centum' and means per hundred; divided by 100 (think of the French word for 100: cent). For example, 7% is 7 divided by 100:

$$7\% = \frac{7}{100} = 0.07$$

And 12% of 50 means:

$$\frac{12}{100} \times 50 = 0.12 \times 50 = 6$$

And in general:

THEOREM 6.1

Percentages

p% of x equals

$$\frac{p}{100} \times x \tag{6.2}$$

Increase and Decrease
In the following two examples we will illustrate the result of a 5% increase of €200, and a 10% decrease of €200.

Example 6.3 Percentage Increase

If €200 increases by 5% then we have 5% of €200 more than before, hence we add the 5% of €200 to the 100% of €200 we already have. The result is 105% of €200, which is €210:

$$\begin{array}{rl} 100\% \times €200 = \quad 1 \times €200 = €200 & \\ 5\% \times €200 = 0.05 \times €200 = €10 & + \\ \hline 105\% \times €200 = 1.05 \times €200 = €210 & \end{array}$$

Hence if 200 increases by 5% the result is:

$$\left(1+\frac{5}{100}\right)\times 200 = 1.05 \times 200$$

Example 6.4 Percentage Decrease

Now, if €200 decreases by 10% then we have 10% less of €200 than before, hence we subtract the 10% of €200 from the 100% of €200 we had. The result is 90% of €200, which is €180:

$$\begin{array}{rl} 100\% \times €200 = \quad 1 \times €200 = €200 & \\ 10\% \times €200 = 0.1 \times €200 = €20 & - \\ \hline 90\% \times €200 = 0.9 \times €200 = €180 & \end{array}$$

Hence if 200 decreases by 10% the result is:

$$\left(1-\frac{10}{100}\right)\times 200 = 0.9 \times 200$$

In general we have the following:

THEOREM 6.2

An *increase* of x by p% results in:

$$\left(1+\frac{p}{100}\right)\times x,$$

whereas a *decrease* of x by p% results in:

$$\left(1-\frac{p}{100}\right)\times x$$

Composition of Increases and Decreases

An increase or decrease may be followed by another increase or decrease. Let us show an example of what happens if €400 increases by 25%, and then decreases by 25%.

Example 6.5 Composition of Increases and Decreases

Suppose we have €400 which increases by 25% in the first year, and then decreases by 25% in the second year. We will show that the overall percentage by which the amount decreases over these two years equals 6.25%.

When the €400 increases by 25% the result is:
$1.25 \times €400 = €500$

When this €500 decreases by 25% the result is:
$0.75 \times €500 = €375$

We have found €375 by first multiplying €400 by 1.25, and then by 0.75. Since $1.25 \times 0.75 = 0.9375$ the net effect is a multiplication by 0.9375:
$€400 \times 1.25 \times 0.75 = €400 \times 0.9375 = €375$

Theorem 6.2 now implies that the factor 0.9375 is equivalent to a percentage decrease of 6.25%.

Beware of the common misconception that:

$€400 + 25\% - 25\% = €400$

6

6.3 Interest

As consumers, we sometimes make speculations regarding our buying decisions, so our consumer credit is based on the idea of us buying something now or in the near future. 'And people's actions are influenced by their expectations. People respond not just to what is happening now (such as change in price), but to what they anticipate will happen in the future' (Sloman & Wride, 2009).

'A bank savings account is based on the opposite idea: saving money now and receiving Interest on that money from the bank. Or borrowing money from the bank and paying back the money with some interest' (Slavin, 2007).

In this section we will discuss the concepts of simple and compounded interest.

Example 6.6 Simple Interest

If a savings account contains €500 and the yearly 4% interest it receives is withdrawn from this account at the moment it is credited, then the balance of this account stays at €500 forever, and the $4\% \times €500 = €20$ yielded each year can be seen as the gain of this account. So after 10 years this account still contains €500, and the owner earned $10 \times €20 = €200$.

Example 6.7 Compounded Interest

If a savings account contains €500 and the yearly 4% interest it receives is reinvested into the same account, then the balance of this account increases by 4% per year:
The balance of this account at the beginning is
€500

After 1 year the balance has increased by 4%:
$1.04 \times €500 = €520$

After 2 years this balance has increased by 4% again, so the balance is:
$1.04 \times €520 = €540.80$

Since $520 = 1.04 \times 500$, according to Example 6.5 this €540.80 is equal to:
$1.04 \times 1.04 \times €500 = (1.04)^2 \times €500$

Continuing, we find that the balance in this account after n years equals
$€500 \times (1.04)^n$

6.4 Time Value of Money

6

The previous example naturally implies the concept of time value of money. Whenever you put your money into a bank account, its value will increase due to compounded interest.

Example 6.8 Future Value

Suppose you deposit €500 in a bank right now, at the beginning of year 0. The interest rate equals $i = 5\% = 0.05$. How much is your investment worth in 4 years' time?

FIGURE 6.1 Future value

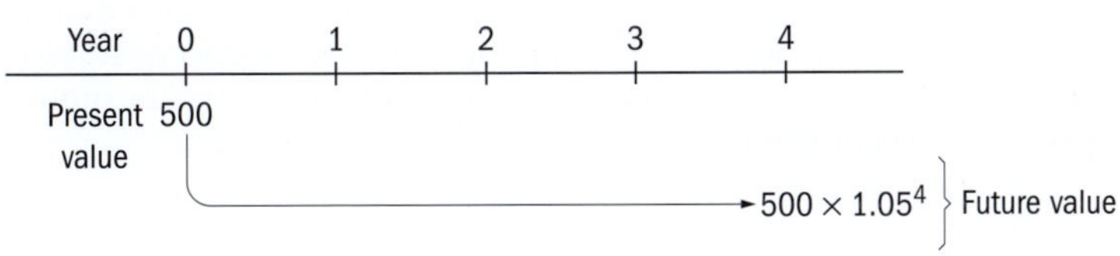

On this timeline (figure 6.1) we can see that a deposit was made at the beginning of year 0. This deposit has a present value of €500. The future value of this investment is $500 \times (1.05)^4$, because the €500 stays in the bank for 4 years, hence it receives 4 times the interest on interest.

In general we have:

THEOREM 6.3

Future Value

The future value FV of an amount of money with present value PV after n years receiving an interest rate of i equals

$$FV = PV \times (1+i)^n \qquad (6.3)$$

Compare this theorem to Theorem 6.2. The term $(1+i)^n$ is called 'future value interest factor' and is often denoted by $FVIF$ or $S_{n\angle i}$. These values are tabulated, or can be calculated. It is a good habit to use at least 3 decimals during the calculation, and to round off the final result to the nearest cent.

6.5 Depreciation of an Asset

In business the term depreciation refers to the allocation of the cost of an asset within several periods of time. A machine, a vehicle or a TV will usually last for more than one year.

6

Definition 6.2 Depreciation
Depreciation is a systematic and rational process of distributing the cost of tangible assets over the lifetime of the asset.

There are several models for depreciating an asset. In this book we will focus only on the two main methods: the straight line (SL) method and the declining balance (DB) method.

6.5.1 Straight Line Method (SL)

This method is the easiest and the most used method in business. It is computed by writing off capital investment linearly over a certain period of time (n). That means taking the purchase price (cost) and subtracting the salvage value (the estimated value that asset will have upon its sale at the end of its useful life), and dividing by the total quantity of productivity years that the asset can be expected to benefit the company. For C as the cost of an asset, S as the salvage value and n as the useful life time in time (e.g. years, months), the depreciation D is given by:

$$D = \frac{C-S}{n}$$

Example 6.9

An asset that cost €25,000, has a salvage value of €5,000 after 5 years. The annual depreciation is:

$$D = \frac{25{,}000 - 5{,}000}{5} = 4{,}000$$

So the asset depreciates €4,000 per year.

6.5.2 Declining Balance Method (DB)

Conversely to the straight line method, the declining balance method assumes a constant depreciation rate i per year. The declining balance is the opposite of compound interest, since:

- compound interest increases the initial value each year by a given percentage (see Example 6.7 and Theorem 6.3)
- depreciation decreases the initial value each year by a given percentage

THEOREM 6.4

The salvage value using the declining balance is equal to

$$S = C(1 - i)^n$$

Example 6.10

Find the salvage value S and the depreciation D of a computer after $n = 10$ months if its price when new was $C = €1{,}500$ and it depreciates by $i = 5\%$ per month. The value equals:

$$S = 1{,}500 \times (1.05)^{10}$$
$$= 898.11$$

So the depreciation D now is:

$$D = C - S$$
$$= 1{,}500 - 898.11$$
$$= 601.89$$

6.5.3 Present and Future Value

Finance is all about time and risk. Present Value and Future Value analysis helps to understand that risk.

Example 6.11 Present Value

Another problem we can solve is how much to put in a bank account right now in order to have €500 in that account (with the same interest rate of $i = 5\% = 0.05$) after 4 years, i.e. what is the present value PV of €500?

FIGURE 6.2 Present value

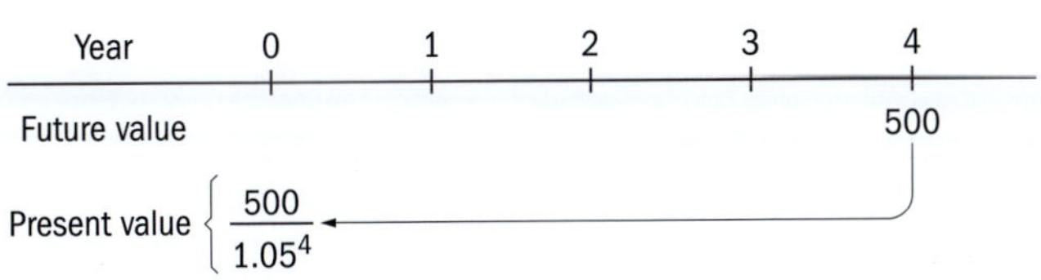

On this time line (figure 6.2) we see that in order to have a future value FV of €500 a deposit of $PV = \dfrac{€500}{(1.05)^4}$ must be made right now, at

$t = 0$. This is because, if you put an amount PV in the bank today, which is worth $FV = €500$ after 4 years, this amount PV must satisfy:
$PV \times (1.05)^4 = 500$

This equation has the solution:

$$PV = \frac{500}{(1.05)^4} = 500 \times (1.05)^{-4} = 411.35$$

This idea can be generalized to the following theorem:

THEOREM 6.5

Present Value

An amount of money which has a future value FV, n years in the future has a present value PV of:

$$PV = \frac{FV}{(1+i)^n} \qquad (6.4)$$

$$= FV(1+i)^{-n}$$

6

The term $(1 + i)^{-n}$ is called 'present value interest factor' and is often denoted by $PVIF$, or $A_{n\angle i}$. These values can be calculated, or can be found in financial tables.

Example 6.12 Duration

An amount of €50,000 was put in a savings account with an interest rate of 2.5% per year. At present the balance of this account equals €76,080.91. No other deposits have been made. How long did this amount stay in the bank?

In order to find the answer, we should solve the equation:
$50{,}000 \times (1.025)^n = 76{,}080.91$

Division by 50,000 simplifies this equation to:
$(1.025)^n = 1.52162$

Taking logarithms and using the rules for calculations with logarithms, equation 3.26 yields:
$n \log(1.025) = \log(1.52162)$

which has the solution:

$$n = \frac{\log(1.52162)}{\log(1.025)} = 17$$

So, if you put €50,000 in a savings account with an interest rate of 2.5%, and wait 17 years, then the value will have increased to €76,080.91.

6.6 Nominal and Effective Interest

At this stage, students should be aware of the difference between Nominal and Effective interest rates. Nominal occurs once per year and Effective occurs and is compounding more than once per year.

Example 6.13 Nominal and Effective Interest

For overdrafts, the ABM bank charges a nominal interest of $j_{12} = 14.1\%$ per year, compounded monthly. What does that mean? This means that each month this bank charges an effective interest i of:

$$i = \frac{14.1\%}{12} = 1.175\% = 0.01175$$

where 12 is the number of months per year.

Suppose you overdrew €500 for the duration of one year. Your debt will have then increased to an amount of:

$$500 \times \left(1 + \frac{0.141}{12}\right)^{12} = 500 \times (1.01175)^{12} = 500 \times 1.1505$$

Hence the effective interest i per year equals:
$1.1505 - 1 = 0.1505 = 15.05\%$

This is more than the nominal interest of 14.1%.

When borrowing money, the nominal interest may come into play, because more interest is charged than appears at first sight.

EXTEND YOUR KNOWLEDGE
That more interest is charged when using nominal interest is a consequence of the Taylor expansion of $\left(1 + \frac{x}{n}\right)^n$: For integer n we have

$$\left(1 + \frac{x}{n}\right)^n = \sum_{i=0}^{n} \binom{n}{i} \left(\frac{x}{n}\right)^i = 1 + x + \binom{n}{2}\left(\frac{x}{n}\right)^2 + \ldots + \left(\frac{x}{n}\right)^n$$

This implies that

$$\left(1 + \frac{x}{n}\right)^n \geq 1 + x \text{ for } x \geq 0$$

The nominal interest can be calculated for various periods of time. Here are some examples:

- A nominal interest rate of $j_{52} = 14.1\%$ per year compounded weekly equals an effective weekly interest rate of $i = \frac{14.1\%}{52} = 0.271\%$. After 1 year, a debt of €500 has then increased to:

$$500 \times \left(1 + \frac{0.141}{52}\right)^{52} = 500 \times (1.00271)^{52} = 500 \times 1.1512$$

Hence the effective yearly interest rate i equals $1.1512 - 1 = 0.1512 = 15.12\%$.

- A nominal interest rate of $j_{365} = 14.1\%$ per year compounded daily equals an effective daily interest rate of $i = \frac{14.1\%}{365} = 0.039\%$. After 1 year, a debt of €500 has then increased to:

$$500 \times \left(1 + \frac{0.141}{365}\right)^{365} = 500 \times (1.00039)^{365} = 500 \times 1.1514$$

These periods of time can be made arbitrarily small (hours, minutes, seconds, etc.), and taken to the other extreme, we have continuous compounding:
When a nominal interest rate of 14.1% per year is compounded continuously a debt of €500 has increased in one year to a debt of:

$$500 \times e^{0.141} = 500 \times 1.1514$$

where $e \approx \sim 2.71828128$, see section 3.6, between Theorem 3.4 and Example 3.6.
Hence the effective yearly interest rate i equals:

$$1.1514 - 1 = 0.1514 = 15.14\%$$

These calculations can be generalized using the following theorem.

THEOREM 6.6

Nominal Interest

After 1 year, a debt of PV increases to a debt FV of:

$$FV = PV \times \left(1 + \frac{j_m}{m}\right)$$

when the nominal yearly interest j_m is compounded m times a year.

After n years a debt of PV increases to a debt FV of:

$$FV = PV \times \left(1 + \frac{j_m}{m}\right)^{nm}$$

when the nominal yearly interest j_m is compounded m times a year, for t number of years.

After 1 year a debt of PV increases to a debt FV of:

$$FV = PV \times e^{j}$$

when the nominal yearly interest j is compounded continuously.

After n years a debt of PV increases to a debt FV of:

$$FV = PV \times e^{nj}$$

when the nominal yearly interest j is compounded continuously.

EXTEND YOUR KNOWLEDGE

The reason for the fact that $\left(1+\frac{j_m}{m}\right)^{nm}$ becomes e^{nj} when the interest deposits are made continuously is because:

$$\left(1+\frac{x}{n}\right)^n \rightarrow e^x \text{ as } n \rightarrow \infty$$

The second equation of theorem 6.6 can be exemplified using the following calculation:

Example 6.14

Suppose you overdrew €500 against a nominal interest rate of $j_{12} = 14.1\%$ compounded monthly. Then your debt will have increased to:

$$FV = 500 \times \left(1+\frac{0.141}{12}\right)^{12}$$

after 1 year.
And to a debt of

$$FV = 500 \times \left(1+\frac{0.141}{12}\right)^{24}$$

after 2 years.

6.7 The Future Value of a Sequence of Payments

What will the balance of your savings account be after 3 years when you invest $PMT_0 = €100$ now (in year 0), $PMT_1 = €150$ at the beginning of the next year (in year 1), and $PMT_2 = €300$ at the beginning of the second year? The interest rate equals $i = 6\% = 0.06$. Let us order this information on a timeline first (figure 6.3).

FIGURE 6.3 Future value of payments

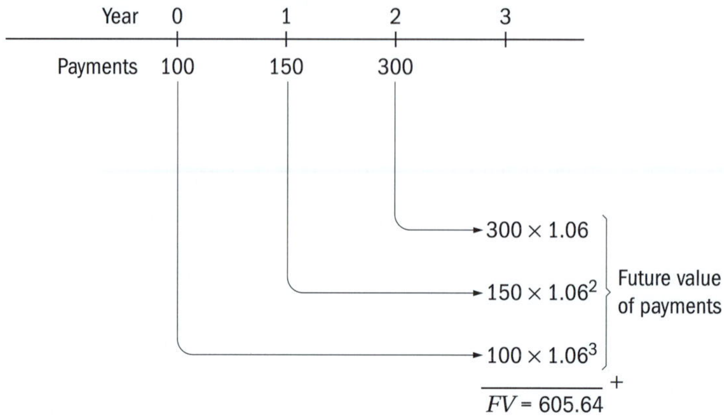

6

The €100 stays in the bank account for 3 years, therefore its future value is $100 \times (1.06)^3$. The €150 stays in the bank for 2 years, therefore its future value is $150 \times (1.06)^2$. And the €300 stays in the bank account for 1 year, therefore its future value is 300×1.06. The balance *FV* in the savings account after these 3 years therefore equals the sum of all the future values of the payments:

$$\begin{aligned} FV &= \sum_{k=0}^{n} PMT_k(1+i)^{n-k} \\ &= 100 \times (1.06)^3 + 150 \times (1.06)^2 + 300 \times 1.06 \\ &= 119.10 + 168.54 + 318 \\ &= 605.64 \end{aligned}$$

This leads to the following theorem:

THEOREM 6.7

The future value *FV* of a sequence of payments $PMT_0, PMT_1, \ldots, PMT_n$ on a savings account against an interest rate of i equals:

$$FV = \sum_{k=0}^{n} PMT_k(1+i)^{n-k} \qquad (6.5)$$

Suppose you have to decide whether or not to invest in a project. This investment costs €100,000, and the returns per year over the next 4 years would be €25,000, €27,000, €32,000, and €34,000. The sum of these 4 returns, €118,000, is more than the initial investment. However, these payments are in the future, and the interest is not accounted for. Is it wise to invest in this project? The bank advises an interest rate of $i = 6\% = 0.06$. The present values, or cash flows, are equal to:

$$PV = \frac{PMT}{(1+i)^k}$$

So the present value of the first payment equals $\frac{€25{,}000}{1.06} = €23{,}584.91$, the present value of the second payment equals $\frac{€27{,}000}{(1.06)^2} = €24{,}029.90$, etc.

In order to analyze this problem, we must first put the information on a time line (figure 6.4).

FIGURE 6.4 Net present value

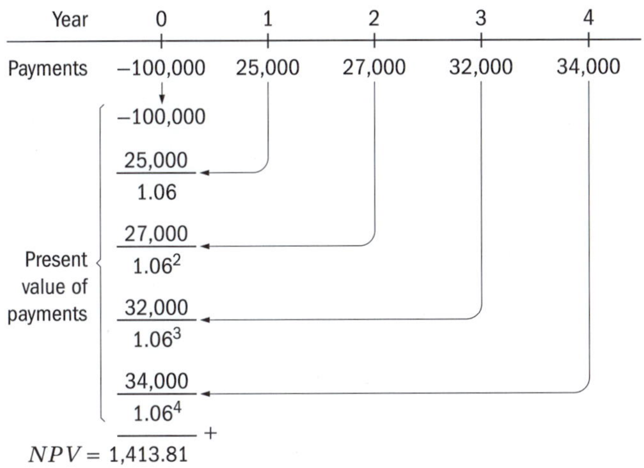

The net present value of this project, *NPV*, equals the sum of all the cash flows:

$$
\begin{aligned}
NPV &= \sum_{k=0}^{n} \frac{PMT_k}{(1+i)^k} \\
&= -100{,}000 + \frac{25{,}000}{1.06} + \frac{27{,}000}{(1.06)^2} + \frac{32{,}000}{(1.06)^3} + \frac{34{,}000}{(1.06)^4} \\
&= -100{,}000 + 23{,}584.91 + 24{,}029.90 + 26{,}867.82 + 26{,}931.18 \\
&= 1{,}413.81
\end{aligned}
$$

This means that this project returns €1,413.81 more than investing the same €100,000 in a bank account. Since this net present value is positive, this investment would add value to the investor. It is therefore advisable to accept this project. If it were negative it would be preferable to invest the same amount of money, the €100,000, in a bank account against the interest rate of $i = 6\%$.

THEOREM 6.8

Net Present Value

The net present value *NPV* of a sequence of payments $PMT_0, PMT_1, \ldots, PMT_n$ equals the sum of all the returns which are corrected for interest:

$$NPV = \sum_{k=0}^{n} \frac{PMT_k}{(1+i)^k} \qquad (6.6)$$

Summary

6

- $\sum_{i=1}^{n} x_i = x_1 + x_2 + \ldots + x_n$
- p% of x is:

  $\frac{p}{100} \times x$

- If x increases by p% the result is:

  $\left(1 + \frac{p}{100}\right) \times x$

- If x decreases by p% the result is:

  $\left(1 - \frac{p}{100}\right) \times x$

- An increase of p% followed by a decrease of q% results in the factor:

  $\left(1 + \frac{p}{100}\right)\left(1 - \frac{q}{100}\right)$

- Simple interest: the amount of money received as interest is immediately withdrawn from the account.
- Compounded interest: the amount of money received as interest stays in the same account.
- When dealing with time value of money, always draw a time line first.
- The future value FV of an amount of money with present value PV is:

  $FV = PV(1+i)^n$

  after n years when the interest rate is i.
- The present value PV of an amount of money with future value FV n years in the future is:

  $PV = \frac{FV}{(1+i)^n} = FV(1+i)^{-n}$

  when the interest rate is i.
- After n years a debt of PV increases to a debt FV of

  $FV = PV \times \left(1 + \frac{j_m}{m}\right)$

  when the nominal yearly interest j_m is compounded m times a year.
- After n years a debt of PV increases to a debt FV of

  $FV = PV \times e^{nj}$

  when the nominal yearly interest j is compounded continuously.
- The future value FV of a sequence of payments PMT_0, PMT_1, ..., PMT_n on a savings account against an interest rate of i equals:

  $FV = \sum_{k=0}^{n} PMT_k(1+i)^{n-k}$

- The net present value NPV of a sequence of payments PMT_0, PMT_1, ..., PMT_n equals the sum of all the returns which are corrected for interest:

  $NPV = \sum_{k=0}^{n} \frac{PMT_k}{(1+i)^k}$

Exercises

6.1 Write without the summation sign:

a $\sum_{i=1}^{4} i$

b $\sum_{i=1}^{3} 2i$

c $\sum_{k=1}^{5} k^2$

d $\sum_{i=1}^{4} 2^i$

e $\sum_{i=1}^{3} x_i$

f $\sum_{i=1}^{4} x_i$

g $\sum_{i=1}^{3} a_i$

h $\sum_{k=1}^{5} x_k$

i $\sum_{i=3}^{7} x_i$

j $\sum_{k=3}^{7} (2k^2 + 1)$

k $\sum_{i=1}^{3} (13 - 2i^2)$

l $\sum_{k=100}^{105} (k - 100)$

6.2 Set
$x_1 = 3, x_2 = 4, x_3 = 7, x_4 = 6, x_5 = 9, x_6 = 1,$
$y_1 = 13, y_2 = 5, y_3 = 11, y_4 = 43, y_5 = 7, y_6 = 24$

Calculate:

a $\sum_{i=1}^{6} x_i$

b $\sum_{i=3}^{6} x_i$

c $\sum_{i=5}^{6} x_i^2$

d $\frac{1}{6}\sum_{k=1}^{6} x_k$

e $\sum_{i=2}^{5} (x_i - y_i)$

f $\sum_{i=3}^{6} (x_i^2 - 2y_i)$

g* $\sum_{i=1}^{3} x_k$

h* $\sum_{k=1}^{5} (x_k - y_2)$

i $\sum_{k=1}^{6} (x_k - 6)$

j $\sum_{k=1}^{6} (x_k - 6)^2$

k $\frac{1}{5}\sum_{k=1}^{6} (x_k - 6)^2$

6.3 Write with a summation sign:

a $1 + 2 + \ldots + 10$
b $3 + 4 + 5$
c $3 + 6 + 9 + 12 + 15$
d $12 + 15 + 18 + 21 + 24$
e* $-2 - 1 + 0 + 1 + 2 + 3$
f* $1 + 2 + 4 + 8 + 16 + 32$
g* $2 + 5 + 8 + 11 + 14$
h* $1 - 3 + 9 - 27 + 81 - 243$

6.4 Write with a summation sign:

a $x_1 + x_2 + \ldots + x_{10}$
b $x_3 + x_4 + x_5$
c $x_1^2 + x_2^2 + x_3^2 + x_4^2$
d $x_2^2 - 1 + x_3^2 - 1 + x_4^2 - 1$

6.5 Calculate

a 1% of 100
b 5% of 100
c 5% of 50
d 50% of 5
e 27% of 27
f 68.73% of 35
g 23.87% of 21.72
h 99.99% of 12,381.67
i 45.67% of 89.01

6.6 Calculate

a 110% of 100
b 555% of 100
c 155% of 50
d 150% of 5
e 217% of 27
f 768.73% of 35
g 823.87% of 21.72
h 100.01% of 12,381.67
i 345.67% of 89.01

6.7 Find the single percentage increase or decrease equivalent to a 20% increase followed by a 25% increase.

6.8 Find the single percentage increase or decrease equivalent to a 30% increase followed by a 35% decrease.

6.9 A couch costs €1,088 including 21.0% V.A.T. Find the new price if the V.A.T. is reduced to 8%.

6.10 A bank offers a return of 4.5% interest compounded annually. Find the future value of a principal of €7,000 after five years. What is the overall percentage rise over this period?

6.11 A piece of machinery depreciates in value by 15% per year. Determine its value in three years' time if its current value is €10,000.

6.12 You know that your favorite car will cost €25,000 in 5 years' time. You wish to invest €21,000 now. What rate of interest would you need to make sure you can buy that car 5 years from now?

6.13 Nikol inherits €99,000. She decides to invest this money in a bank. The bank tells her that her money will increase in value with a fixed yearly percentage i. After exactly one year the value of her money is €105,000. Calculate this fixed yearly percentage i (required accuracy: to the nearest hundredth of a percent).

6.14 If a piece of machinery depreciates continuously at an annual rate of 10%, how many years will it take for the value of the machinery to halve?

6.15 How much is €100 in a savings account worth in 10 years' time if the yearly interest rate is

a 2.1%
b 4.2%
c* How much interest is received in **a** and **b**?

6.16 What amount of money should be deposited in a savings account in order to have €100 in 10 years' time if the interest rate is

a 2.1%
b 4.2%

6.17 Assume an interest rate of 6%. How much would you have to put in a savings account in order to have €15,000

a next year?
b 2 years from now?
c 10 years from now?

6.18 How many years does it take for €1,000 to double if the yearly interest rate is 3%?

6.19

a Which earns more: €1,000 in a savings account for 10 years with an interest rate of 6%, or €1000 in a savings account for 20 years with an interest rate of 3%?
b* Suppose a savings account contains an amount $PV > 0$. Is it more profitable to have an interest rate of i for $2n$ years or to have an interest rate of $2i$ for n years?

6.20 Five years ago Dr. Lu bought a house for €400,000. Its value increased according to the compound interest formula at an annual rate of 8%. What is the current value of Dr. Lu's house?

6.21 The revenue of a very stable company was €25,000,000 in 2006. In 2011 its revenue was €30,000,000. Calculate the average percentage growth per year during that period.

6.22 Find the effective yearly rate of interest corresponding to a nominal rate of 20% compounded:

a annually
b semi-annually
c monthly
d weekly
e daily
f continuously

6.23 A principal of €1 is invested at 20% interest for one year. Determine its future value if the interest is compounded

a annually
b semi-annually
c monthly
d weekly
e daily
f continuously

6.24 A principal of €1 is invested at 20% interest for five years. Determine its future value if the interest is compounded:
- **a** annually
- **b** semi-annually
- **c** monthly
- **d** weekly
- **e** daily
- **f** continuously

6.25 A principal of €8,000 is invested at 20% interest for five years. Determine its future value if the interest is compounded
- **a** annually
- **b** semi-annually
- **c** monthly
- **d** weekly
- **e** daily
- **f** continuously

6.26 A principal of €3,000 is invested at 3.5% interest for five years. Determine its future value if the interest is compounded:
- **a** annually
- **b** semi-annually
- **c** monthly
- **d** weekly
- **e** daily
- **f** continuously

6.27* Suppose that this morning a jellyfish lying on the beach consisted of 99% water. In the evening, after some sunbathing, it consists of 98% water. By how much did its size decrease? Assume that only the water vaporized.

6.28 What will be the balance of a savings account after 3 years if €50 was deposited at the beginning of this year, €250 will be deposited at the beginning of the next year and €170 another year later? Assume an interest rate of 2.6%.

6.29 An investment in a project costs €1,000,000. You will be paid €250,000 after 1 year, €300,000 after 2 years, €350,000 after 3 years, and €400,000 after 4 years. The bank advices a discount rate of 5.3%. What is the net present value of this project?

7
Arithmetic and Geometric Series

At the end of this chapter you should be able to:
- understand the difference between arithmetic and geometric series
- calculate the number of terms and the sum of arithmetic series
- calculate the number of terms and the sum of geometric series
- apply real-life applications of the series

To understand the time value of an investment, which is the present value and the future value of money, you first need to understand how to use series, especially the geometric series. This will lead you to understand the concepts of annuity and amortization, which will be discussed in chapter 8.

Gauss and the series

Carl Friedrich Gauss was the inventor of normal distribution, and developer of the geometry of curved spaces which was later used by Einstein. The story goes that when Gauss was only 10 years old, his teacher asked his pupils to add up all the integers between 1 and 100. Gauss almost immediately came up with the correct answer, 5,050. How did he do that?

Carl Friedrich Gauss (1777–1855)

7.1 Series

A series is the sum of a sequence, or progression, of numbers. For example, from the sequence of numbers 1, 2, 3, 4 the sum equals:

$$\sum_{i=1}^{4} i = 1+2+3+4=10$$

We will consider two kinds of series. One in which the *difference* between two consecutive terms is constant, such as

$$\sum_{i=1}^{10} 5i = 5+10+15+20+25+30+35+40+45+50$$

This is called an *arithmetic* series.
And the other where the *ratio* of two consecutive terms is constant, such as

$$\sum_{i=0}^{9} 2^i = 1+2+4+8+16+32+64+128+256+512$$

This is called a *geometric* series.

7.2 Arithmetic Series

An arithmetic series is a summation of a sequence of numbers in which the difference between any two consecutive terms is constant, such as

$1+2+3+4+5+6$ common difference equals 1
$6+5+4+3+2+1$ common difference equals -1
$3+6+9+\ldots+30$ common difference equals 3

In the arithmetic series $S=1+2+\ldots+100$ we recognize that:
- the first term, denoted by t_1, equals 1
- the last term, denoted by t_n, equals 100
- the amount of terms, denoted by n, equals 100
- the common difference of any two consecutive terms, denoted by d, equals 1

When asked to sum up these integers Gauss realized that the order of addition does not matter:

$$\begin{aligned} S &= 1+2+\ldots+100 \\ &= 100+99+\ldots+1 \end{aligned}$$

When he put both of these additions on top of each other he saw that

$$\begin{array}{lcl} S & = & 1 + 2 + \ldots + 99 + 100 \\ \underline{S} & \underline{=} & \underline{100 + 99 + \ldots + 2 + 1 +} \\ S+S & = & 101 + 101 + \ldots + 101 + 101 \end{array}$$

Hence:

$$2S = 100 \times 101$$

Therefore:

$$S = \frac{100 \times 101}{2} = \frac{10{,}100}{2} = 5{,}050$$

Gauss' procedure of finding the sum of an arithmetic series can be generalized for any numbers t_1, t_n, d, and n. The result is the following theorem.

THEOREM 7.1

Arithmetic Series

The sum S of an arithmetic series $\sum_{i=1}^{n} t_i$ is given by

$$S = \frac{n}{2}(t_1 + t_n) \qquad (7.1)$$

where n is the amount of terms, t_1 and t_n the first and the last term of the series respectively.

We will give another example of the use of this theorem.

Example 7.1

Let us find the sum of the arithmetic series:

$$\underbrace{S = 1 + 3 + 5 + 7 + 9 + 11 + 13 + 15 + 17 + 19}_{n = 10 \text{ terms}}$$

We recognize $t_1 = 1$, $t_n = 19$, $d = 2$, and $n = 10$. According to formula (7.1) the sum equals:

$$\begin{aligned} S &= \frac{10}{2}(1 + 19) \\ &= 5 \times 20 \\ &= 100 \end{aligned}$$

Furthermore, there is a formula to calculate t_n, the nth term of an arithmetic series. This may be of use when the number of terms n of an arithmetic series is unknown.

THEOREM 7.2

The *n*th term of an arithmetic series

The *n*th term of an arithmetic series, t_n, is given by

$$t_n = t_1 + (n - 1)d \qquad (7.2)$$

with t_1 the first term of the series, and d the difference between any two consecutive terms of the series.

As an example of the application of both Theorem 7.1 and Theorem 7.2 we shall consider the following example.

Example 7.2

What is the sum of the arithmetic series $196+193+\ldots+4$?
We see that $t_1=196$, $t_n=4$, $d=-3$, but n is cumbersome to determine without formula (7.2).

$$4=196+(n-1)\times(-3)$$
$$-192=-3(n-1)$$
$$\frac{-192}{-3}=n-1$$
$$64=n-1$$

Hence:
$n=65$ terms
Now we can easily plug all of these numbers into formula 7.1 of the sum of an arithmetic series:

$$S=\frac{65}{2}(196+4)$$
$$=\frac{65}{2}\times 200$$
$$=65\times 100$$
$$=6{,}500$$

7

An Application of Arithmetic Progression

One of the authors running a half-marathon

After a knee surgery, your coach advises you to resume your jogging program gradually. He suggests you go jogging for 12 minutes each day for the first week. Each week thereafter, he suggests you to increase that time by 6 minutes per day. How many weeks will it be before you are up to jogging 60 minutes per day (source: www.regentsprep.org)?

This is an arithmetic progression:

$$t_1=12$$
$$t_n=60$$
$$d=6$$

$$t_n=t_1+(n-1)d$$
$$60=12+(n-1)\times 6$$
$$60=12+6n-6$$
$$54=6n$$
$$n=9$$

So it will take 9 weeks before you are up to jogging 60 minutes per day.

7.3 Geometric Series

A geometric series is a summation of a sequence of numbers in which the ratio between two consecutive terms is constant, such as:

$1+2+4+8+16+32$ common ratio equals 2

$32+16+8+4+2+1$ common ratio equals $\frac{1}{2}$

$\frac{1}{3}+\frac{1}{9}+\frac{1}{27}+\frac{1}{81}+\frac{1}{243}$ common ratio equals $\frac{1}{3}$

The next chapter on Financial Mathematics relies heavily on the application of geometric series. We will explain the geometric series using the following example.

Example 7.3

Let us add the numbers of the sequence 1, 3, 9, 27, 81. First we denote the sum of these integers 1, 3, 9, 27, 81 by S. We see that every next term is the preceding term multiplied by 3, hence it is a geometric series with a common ratio of 3. Now, if we subtract S from $3S$ we get:

$$\begin{array}{lll} 3S & = & \quad 3+9+27+81+243 \\ S & = & 1+3+9+27+81 \quad - \\ \hline 3S-S & = & -1 \qquad\qquad\qquad +243 \end{array}$$

Hence
$3S-S=S(3-1)=243-1$

Therefore
$S=\frac{243-1}{3-1}=121$

In this geometric series $S=1+3+\ldots+81$ we can recognize that:
- the first term, denoted by a_1, equals 1
- the last term, denoted by a_n, equals 81
- the amount of terms, denoted by n, equals 5
- the common ratio of any two consecutive terms, denoted by r, equals 3

The procedure of finding the sum of a geometric series can be generalized for any a_1, a_n, r, and n. The result is the following theorem:

THEOREM 7.3

Geometric Series

The sum S of a geometric series $\sum_{i=1}^{n} a_i$ is given by:

$$S=a_1\left(\frac{r^n-1}{r-1}\right) \qquad (7.3)$$

where a_1 is the first term of the sequence, n the number of terms, and r the common ratio.

Example 7.4

In Example 7.3 ($S=1+3+\dots+81$) we have $a_1=1$, $a_n=81$, $n=5$, and $r=3$. Hence according to formula (7.3) the sum S equals:

$$S=1\times\frac{3^5-1}{3-1}$$
$$=\frac{243-1}{2}$$
$$=121$$

Example 7.5

Using another example: the series

$$S=\sum_{i=1}^{6}\left(\frac{1}{2}\right)^i=\underbrace{\frac{1}{2}+\frac{1}{4}+\frac{1}{8}+\frac{1}{16}+\frac{1}{32}+\frac{1}{64}}_{n=6\text{ terms}}$$

has as the first term $a_1=\frac{1}{2}$, common ratio $r=\frac{1}{2}$ and $n=6$ terms.

Using formula (7.3) we see that this series is equal to:

$$S=\frac{1}{2}\times\frac{\left(\frac{1}{2}\right)^6-1}{\frac{1}{2}-1}$$
$$=\frac{1}{2}\times\frac{\frac{1}{64}-1}{\left(-\frac{1}{2}\right)}$$
$$=\frac{63}{64}$$

7

Just as formula (7.2) is for calculating the nth term of an arithmetic series, there is a formula to calculate a_n, the nth term of a geometric series too. This may be of use when the number of terms n of a geometric series is unknown.

EXTEND YOUR KNOWLEDGE

What happens if we keep on adding terms to this geometric series, e.g., how much is:

$$S=\frac{1}{2}+\frac{1}{4}+\frac{1}{8}+\dots?$$

The sum S equals

$$S=\frac{1}{2}\times\frac{\lim_{n\to\infty}\left(\frac{1}{2}\right)^n-1}{\frac{1}{2}-1}$$

Because

$$\lim_{n\to\infty}\left(\frac{1}{2}\right)^n=\frac{1}{2}\times\frac{1}{2}\times\ldots=0$$

the sum S equals

$$S=\frac{1}{2}\times\frac{(0-1)}{\left(\frac{1}{2}-1\right)}=\frac{1}{2}\times\frac{(-1)}{\left(-\frac{1}{2}\right)}=1$$

THEOREM 7.4

The nth term of a geometric series

The nth term of a geometric series a_n is given by:

$$a_n=a_1r^{n-1} \tag{7.4}$$

Note the similarities between this and equation (7.2) for the nth term of an arithmetic series.

7

Example 7.6

To show an example of the application of both Theorem 7.3 and Theorem 7.4 we calculate the geometric series:
$S=2+6+\ldots+39{,}366$

This series has $a_1=2$, $r=3$, and $a_n=39{,}366$. With formula (7.4) we can determine the number of terms n:

$$39{,}366=2\times3^{n-1}$$
$$19{,}683=3^{n-1}$$
$$\log(19{,}683)=\log(3^{n-1})$$
$$\log(19{,}683)=(\mathrm{n}-1)\log3$$

Hence

$$n-1=\frac{\log(19{,}683)}{\log3}=9$$

And thus $n=10$.

Now we have all the input for formula (7.3):

$$S=a_a\frac{r^n-1}{r-1}$$
$$=2\times\frac{3^{10}-1}{3-1}$$
$$=2\times\frac{59{,}049-1}{2}$$
$$=59{,}048$$

Summary

- A series is the sum of a sequence of numbers, such as $1+2+3+4$.
- An arithmetic series is a sum of the sequence of numbers in which the difference between any two consecutive terms is constant, such as $2+4+6+8+10$, where the common difference is 2.
- The sum of an arithmetic series is given by formula (7.1):

  $$S=\frac{n}{2}(t_1+t_n)$$

  with n the number of terms, t_1 and t_n the first and the last term respectively.
- The nth term of an arithmetic series is given by formula (7.2):

  $$t_n=t_1+(n-1)d$$

  where d is the difference between two terms.
- A geometric series is the sum of a sequence of numbers in which the ratio between any two consecutive terms is constant, such as $1+3+9+27+81+243$, where the common ratio is 3.
- The sum of a geometric series is given by formula (7.3):

  $$S=a_1\frac{r^n-1}{r-1}$$

  with n the number of terms, a_1 the first term and r the common ratio.
- The nth term a_n of a geometric series is given by formula (7.4):

  $$a_n=a_1r^{n-1}$$

  where r is the common ratio and a_1 the first term.

7

Exercises

7.1 Calculate the following arithmetic series:

a $1+2+3+\ldots+10\,(n=10 \text{ terms})$
b $10+9+8+\ldots+1\,(n=10 \text{ terms})$
c $2+4+6+\ldots+20\,(n=10 \text{ terms})$
d $2+5+8+11+14+17+20+23$
e $(-6)+(-2)+2+6+10+14+18$
f $(-11)+(-9)+(-7)+\ldots+11$

7.2 Calculate:

a $2+3+\ldots+98+99$
b $1+2+\ldots+99+100$
c $0+1+\ldots+100+101$
d $(-1)+0+\ldots+101+102$

7.3 Calculate the following geometric series:

a $2+6+18+\ldots$ (17 terms)
b $\frac{1}{2}+\frac{1}{4}+\frac{1}{8}+\ldots$ (10 terms)
c $3+6+12+\ldots+393{,}216$
d $3{,}125+625+125+\ldots+1$

7.4 Calculate:

a $18+23+28+\ldots+83$
b $2+6+18+\ldots+486$
c $2+6+10+\ldots+486$

7.5 Calculate:

a $\sum_{i=1}^{100} i$
b $\sum_{i=1}^{10} 3+4i$
c* $\sum_{i=1}^{10} 3+4k$
d $\sum_{i=3}^{7} 5^i$
e $\sum_{i=1}^{10} \left(\frac{1}{2}\right)^i$
f $\sum_{i=0}^{10} \left(\frac{1}{2}\right)^i$

7.6* Calculate the square root of the sum of the first 7 odd numbers.

7.7 Consider the following series:

$$\sum_{i=0}^{3} 500 \times (1.03)^i$$

a Is this an arithmetic or a geometric series? Why?
b What is a_1, what is r, and what is n?
c What is the result of this series?

7.8 Consider the following series :

$$500 \times (1.03)^3 + 500 \times (1.03)^2 + 500 \times 1.03 + 500$$

a What is a_1, what is r, and what is n?
b What is the result of this series?
c What is the similarity with exercise 7.7?

7.9 According to legend, the game of chess was invented around 1,500 years ago in India by Sessa ebn Daher. He gave it as a present to his king, Sheram. King Sheram liked this game so much that he let Mr. Daher choose a reward of his own choice. Mr. Daher asked for 1 grain of rice on the first field of the chess board, 2 on the second, 4 on the third, etc. At first, the king was upset about such a small reward. However, the king's accountants soon calculated this reward was not so small at all. How many grains of rice did the king have to give to Mr. Daher? A chess board has 64 fields.

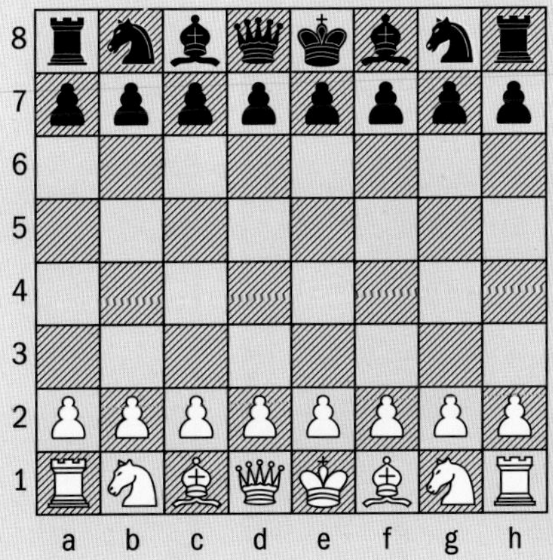

8
Annuity and Amortization

At the end of this chapter you should be able to:
- understand the concept of annuity
- understand different types of annuity
- calculate the present value of an annuity
- calculate the future value of an annuity
- calculate the value of an annuity
- understand the concept of amortization
- calculate the remaining debt

This chapter shows how to deal with a sequence of recurring, equal payments when there is interest involved. We call them *annuities.* There are different kinds of annuities: one in which the largest transaction is being made somewhere in the future, and one in which the largest transaction is being made at the present. An example of the first annuity is a savings account: each period an amount of money is put into a savings account, and somewhere in the future this money can be withdrawn. This withdrawal is the largest transaction, and it lies in the future. An example of the second annuity is an amortization: a large amount of money is borrowed from the bank (at present) and then it is being paid back in smaller amounts of money. Furthermore, the payments can be made at the beginning and at the end of each period. This gives us four kinds of annuities: the largest transaction is at present, or in the future, the payments occur at the beginning or end of the periods:
1 future, end
2 future, beginning
3 present, end
4 present, beginning

Beware: annuities deal with recurring payments. When there is only one payment, use formulas (7.1) and (7.2).

Getting Smart About Annuities

For years, many retirees were content to act as their own pension managers, a complex task that involves making a nest egg last a lifetime. Now, reeling from the stock-market meltdown, many are calling it quits - and buying annuities to do the job for them. Annuities in general have never been popular with many financial advisers. For the most part, the products don't offer the potential for outsized gains. And once you hand over your money to an insurer, you either can't get it back or can do so only by forfeiting at least some of the guarantee you've paid for. Variable annuities, in particular, can be ridiculously complex and loaded with fees and hidden traps.

Source: http://online.wsj.com

8.1 Savings account

This first section explains how the future value of an annuity works. A savings account is a special type of financial product which is called an annuity. An annuity is a recurring set of payments. The calculations in this chapter will turn out to be applications of geometric series. In this book we will restrict ourselves to the 'certain annuity'. Certain means we are sure the parties involved will not cease to exist during the duration of the annuity.
In this section we will find out how to calculate the balance of a savings account after a few yearly deposits of an equal amount against a certain interest rate.

Example 8.1 Savings Account, Payments at the End of the Year

Suppose a deposit of €500 is made at the end of each year. The first deposit is made at the end of year 0 (December 31st at 24:00 hours. which is the same as the beginning of year 1), and the last at the end of year 3. The interest rate is 3%. What is the balance of this savings account after 4 years, that is, just after the last deposit?
We need to draw a timeline first, see figure 8.1.

FIGURE 8.1 End of year savings account payments

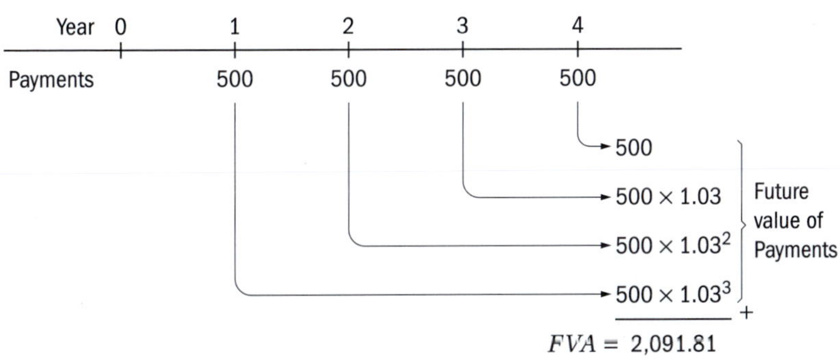

The balance *FVA* (Future Value Annuity) of the savings account after 4 years equals the sum of all the future values of the payments:

$$FVA = 500 \times (1.03)^3 + 500 \times (1.03)^2 + 500 \times 1.03 + 500$$

$$= 500 + 500 \times 1.03 + 500 \times (1.03)^2 + 500 \times (1.03)^3 \qquad (8.1)$$

In the second step the order of addition was reversed.
This is a geometric series with first term $a = 500$, common ratio $r = 1.03$ and $n = 4$ terms. Its sum is:

$$FVA = 500 \times \frac{(1.03)^4 - 1}{1.03 - 1}$$

$$= 500 \times \frac{(1.03)^4 - 1}{0.03}$$

$$= 2{,}091.81$$

8

Compare this calculation with Exercises 7.7 and 7.8. This example can be generalized using the following theorem:

THEOREM 8.1

Future Value of an Ordinary Annuity

A savings account contains

$$FVA = PMT \times \frac{(1+i)^n - 1}{i} \qquad (8.2)$$

where n deposits of PMT are made at the end of every year for n years. Because the payments occur at the *end* of each year, this is an ordinary annuity.

The term $\frac{(1+i)^n - 1}{i}$ is often denoted by *FVIFA* (Future Value Interest Factor Annuity) or by $s_{n\angle i}$ and can be calculated by hand, by using a graphical or financial calculator, or can be found in financial tables.

Explanation

The future value *FVA* of an ordinary annuity can be determined similar to the steps above. First we put the information on a timeline, see figure 8.2.

FIGURE 8.2 Future value of an ordinary annuity

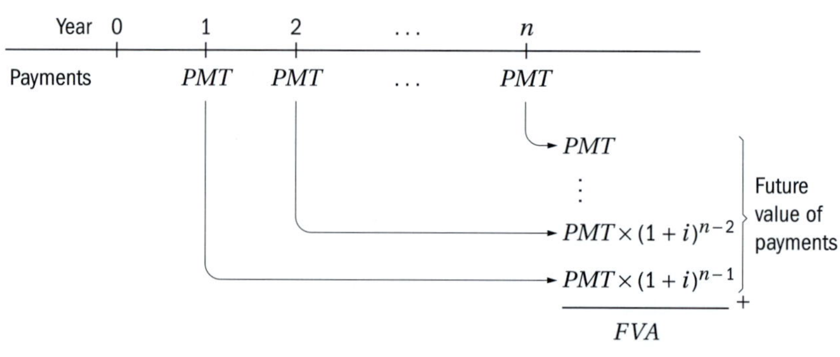

The future value *FVA* of this annuity equals, similar to equation (8.1) above,

$$\begin{aligned} FVA &= PMT \times (1+i)^{n-1} + PMT \times (1+i)^{n-2} + \ldots + PMT \\ &= PMT + PMT \times (1+i) + \ldots + PMT(1+i)^{n-1} \end{aligned}$$

which is a geometric series with first term $a_1 = PMT$, common ratio $r = (1+i)$ and amount of terms n. The result is:

$$FVA = PMT \times \frac{(1+i)^n - 1}{1 + i - 1}$$
$$= PMT \times \frac{(1+i)^n - 1}{i}$$

This is formula (8.2).

8.1.1 Determining the Payments

In this section we will examine the payments of an annuity. Annuities can be paid for a specific period of time such as 30 years or during the annuity owner's lifetime.

Example 8.2

Suppose you want to save $FVA = €2{,}091.81$ in 4 years' time, that is to say, after 4 deposits, while receiving an interest rate of 3%. The payments PMT to be made at the end of each year can be found by solving formula (8.2) for PMT:

$$2{,}091.81 = PMT \times \frac{(1.03)^4 - 1}{0.03}$$
$$= PMT \times 4.183627$$

Solving for PMT gives:

$$PMT = \frac{2{,}091.81}{4.183627}$$
$$= 500$$

This idea can be generalized using the following theorem:

8

THEOREM 8.2

Payments of an Annuity

Finding the payments PMT which should be made in order to have an amount of FVA in a savings account after n years against an interest rate of i can be found by solving equation (8.2):

$$FVA = PMT \times \frac{(1+i) - 1}{i}$$

for PMT.

Example 8.3 Savings Account, Payments at the Beginning of the Year

Now suppose these deposits of €500 are made at the beginning of each year. The first deposit is made right now, at the beginning of year 0, and the last at the beginning of year 3. The interest rate is 3%. What is the balance of this savings account after 4 years? We draw a time line first, see figure 8.3.

FIGURE 8.3 Beginning of year savings account payments

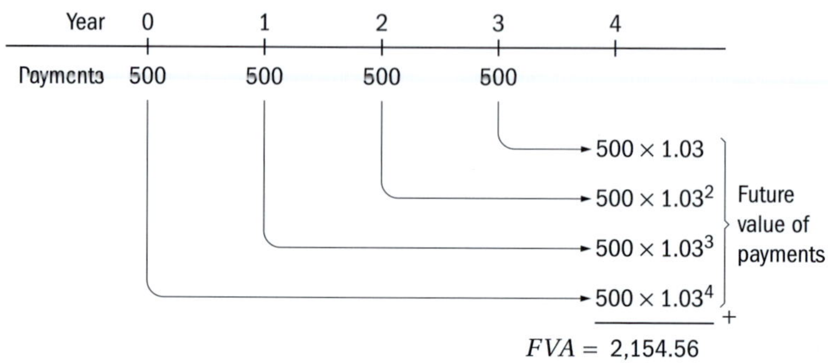

The balance of this savings account after 4 years equals the sum of all the future values of the payments:

$$\begin{aligned} FVA &= 500 \times (1.03)^4 + 500 \times (1.03)^3 + 500 \times (1.03)^2 + 500 \times 1.03 \\ &= 500 \times 1.03 + 500 \times (1.03)^2 + 500 \times (1.03)^3 + 500 \times (1.03)^4 \end{aligned}$$

This is a geometric series with first term $a = 500 \times 1.03$, common ratio $r = 1.03$ and $n = 4$ terms. Its sum is:

$$\begin{aligned} FVA &= 500 \times 1.03 \times \frac{(1.03)^4 - 1}{1.03 - 1} \\ &= 500 \times 1.03 \times \frac{(1.03)^4 - 1}{0.03} \\ &= 1.03 \times 2{,}091.81 \\ &= 2{,}154.56 \end{aligned}$$

This is just 1.03 times more than the first annuity we encountered in this chapter. The reason behind this is that all the payments stay in the bank account one year longer; hence the final balance is 1.03 times bigger. When the payments are made at the beginning of the year we call this an annuity *due*. This example can be generalized using the following theorem:

THEOREM 8.3

Future Value of an Annuity Due

A savings account contains:

$$FVA = PMT \times \frac{(1+i)^n - 1}{i} \times (1+i) \qquad (8.3)$$

when n deposits of PMT are made at the beginning of every year for n years.

Explanation
All the payments of an annuity due stay in the bank 1 year longer, so the final result must be bigger by a factor of $1 + i$.
Remember that payments made at the *end* of the year lead to an ordinary annuity, and payments made at the *beginning* of the year lead to an annuity *due*.

8.1.2 Interest Received

Generally when you purchase an annuity, you earn a certain interest rate for a set period of time. At the end of the period, a new fixed rate becomes effective, based on current interest rates (source: www.prudential.com).

Example 8.4

The amount of interest received on a savings account can be determined very easily: in the previous example the amount of money deposited in the savings account was equal to $4 \times €500 = €2{,}000$. The balance of the savings account after 4 years, *FVA*, is €2,154.56. The difference between the amount deposited and the balance equals $€2{,}154.56 - €2{,}000 = €154.56$. That is the interest received.

In general:

THEOREM 8.4

Interest received on a savings account

Let *FVA* be the balance of a savings account after n payments of *PMT* against an interest rate of i.
The total amount of interest received, R, equals:

$$R = FVA - n \times PMT \qquad (8.4)$$

8

8.2 Present Value of an Annuity

In this section we will show how to calculate the present value of an annuity, and how to calculate the payments per year if the present value of an annuity is known.
Two examples of the application of the present value of an annuity are:
1 an amortization. This is a financial product where an amount of money borrowed from a bank is to be paid back in equal ordinary instalments (payments). A mortgage is an example of an amortization.
2 an annuity. This is a financial product where the buyer of the annuity gets paid out a fixed amount of money for a certain period. Accumulating the prize for a lottery in a certain amount of years is an example of an annuity.

The calculations for an amortization will turn out to be equal to those of an annuity, as we will show in the section on amortization.

8.2.1 Determining the Present Value of an Annuity

In this section we will first show how to determine the present value of an annuity if the yearly payments, the duration, and interest are known. Consider the following example:

Example 8.5

Suppose you want to buy an annuity which pays out $PMT = €15{,}000$ at the end of the year for the next 10 years. How much does that cost now? Let us assume the interest rate i equals $6\% = 0.06$.

FIGURE 8.4 Present value of an annuity

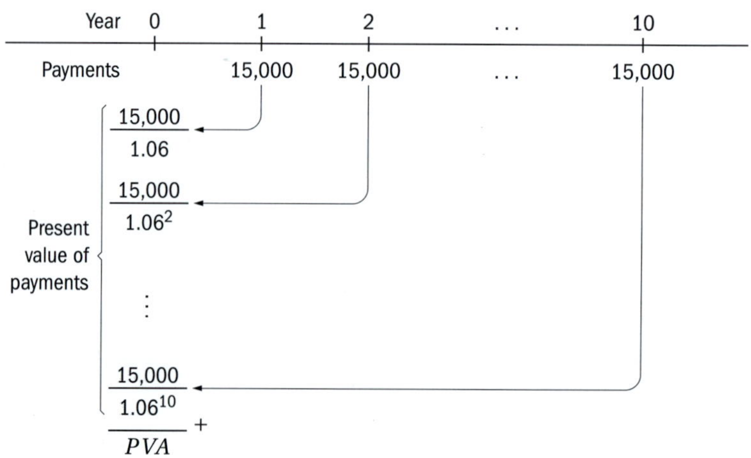

On this timeline (see figure 8.4), we can see that from year 1 up to year 10 you get paid out €15,000 per year. The present value of the €15,000 which is paid out in year 1 equals $\frac{€15{,}000}{1.06}$. Hence in order to receive €15,000 next year you should deposit $\frac{€15{,}000}{1.06}$ today, in year 0. The year after you get paid out another €15,000 which is worth today, $\frac{€15{,}000}{(1.06)^2}$, hence you should also deposit $\frac{€15{,}000}{(1.06)^2}$ today, in year 0. Continuing this argument we see that you should make a total deposit *PVA* of

$$PVA = \frac{15{,}000}{1.06} + \frac{15{,}000}{(1.06)^2} + \ldots + \frac{15{,}000}{(1.06)^{10}} \tag{8.5}$$

in year 0. We recognize this expression as a geometric series with first term $a_1 = \frac{15{,}000}{1.06}$, common ratio $r = \frac{1}{1.06} = (1.06)^{-1}$ and amount of terms $n = 10$. The result is:

$$PVA = \frac{15{,}000}{1.06} \times \frac{\left((1.06)^{-1}\right)^{10} - 1}{(1.06)^{-1} - 1}$$

$$= 15{,}000 \times \frac{\left((1.06)^{-1}\right)^{10} - 1}{1 - 1.06} \qquad (8.6)$$

$$= 15{,}000 \times \underbrace{\frac{1 - (1.06)^{-10}}{0.06}}_{7.3600871}$$

$$= 110{,}401.31$$

So, in order to get paid out $PMT = €15{,}000$ at the end of the year for the next 10 years, you should deposit €110,401.31 in year 0. This deposit is the present value of the annuity, PVA.

This calculation can be generalized such that we can determine the present value of every annuity.

THEOREM 8.5

Present Value of an Ordinary Annuity

The present value PVA of an ordinary annuity, which pays out PMT per year for the next n years, when the yearly interest rate equals i, is

$$PVA = PMT \times \frac{1 - (1 + i)^{-n}}{i} \qquad (8.7)$$

8

The term $\frac{1 - (1 + i)^{-n}}{i}$ is often denoted by $PVIFA$ (Present Value Interest Factor Annuity or $a_{n\angle i}$, and can be found in tables or on a financial calculator. Do compare these formulas (8.7) and (8.5) with formulas (8.1) and (8.2). Since the payments occur at the *end* of the year, this is an ordinary annuity.

Explanation of a present value of an ordinary annuity on a whiteboard

Explanation
The present value *PVA* can be determined in a similar way to the steps above. First we put the information on a timeline (see figure 8.5).

FIGURE 8.5 Present value of an ordinary annuity

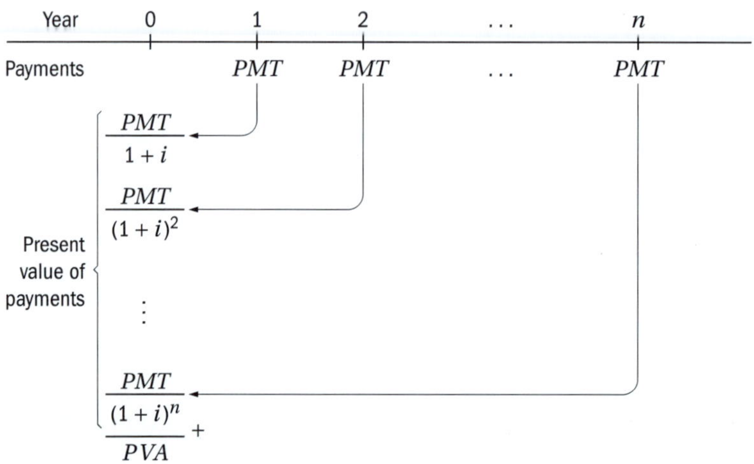

The present value *PVA* of this annuity equals the result below, similar to equation (8.5) above:

$$PVA = \frac{PMT}{1+i} + \frac{PMT}{(1+i)^2} + \ldots + \frac{PMT}{(1+i)^n}$$

Which is a geometric series with first term $a_1 = \frac{PMT}{(1+i)}$, common ratio $r = \frac{1}{(1+i)} = (1+i)^{-1}$ and amount of terms n. The result is:

$$PVA = \frac{PMT}{1+i} \times \frac{(1+i)^{-n}-1}{(1+i)^{-1}-1}$$

$$= PMT \times \frac{(1+i)^{-n}-1}{1-(1+i)}$$

$$= PMT \times \frac{(1+i)^{-n}-1}{-i}$$

$$= PMT \times \frac{1-(1+i)^{-n}}{i}$$

This is formula (8.7). Do compare this with formula (6.6). The difference is that with annuities all payments are equal.

8.2.2 Annuity, Determining the Payments

In this section we will show how to determine the payments of an annuity if its present value, its interest and its duration are known. Consider the following example:

Example 8.6 Lottery Prize Part 1 (Present Value of an Ordinary Annuity)

Imagine you won a prize on the lottery, so today you own a capital of $PVA = €400{,}000$. You would like to buy an annuity which pays out an amount PVA at the end of each year for the following 30 years. The bank offers an interest rate of $i = 6\% = 0.06$. We insert these figures into formula 8.7 and solve for PMT:

$$PVA = PMT \times \frac{1-(1+i)^{-n}}{i}$$

$$400{,}000 = PMT \times \underbrace{\frac{1-(1.06)^{-30}}{0.06}}_{13.764831}$$

$$= PMT \times 13.764831$$

Solving this equation for PMT yields:

$$PMT = \frac{400{,}000}{13.764831}$$

$$= 29{,}059.56 \qquad (8.8)$$

So, the yearly payments equal €29,059.56 if you want to realise your €400,000 in 30 years of equal payments, receiving an interest rate of 6%.

8

This calculation can be generalized such that we can determine the yearly payments of every annuity with present value PVA.

THEOREM 8.6

Payments of an Ordinary Annuity

Let be PVA the present value of an ordinary annuity, i the interest rate, and n the duration of this annuity. Then the payments PMT can be found by solving equation 8.7 for PMT:

$$PVA = PMT \times \frac{1-(1+i)^{-n}}{i}$$

Example 8.7 Present Value of an Annuity Due

The 10 payments of €15,000 in Example 8.5 can also be met at the beginning of the year. On a timeline it will look like this (see figure 8.6).

FIGURE 8.6 Present value of an annuity due

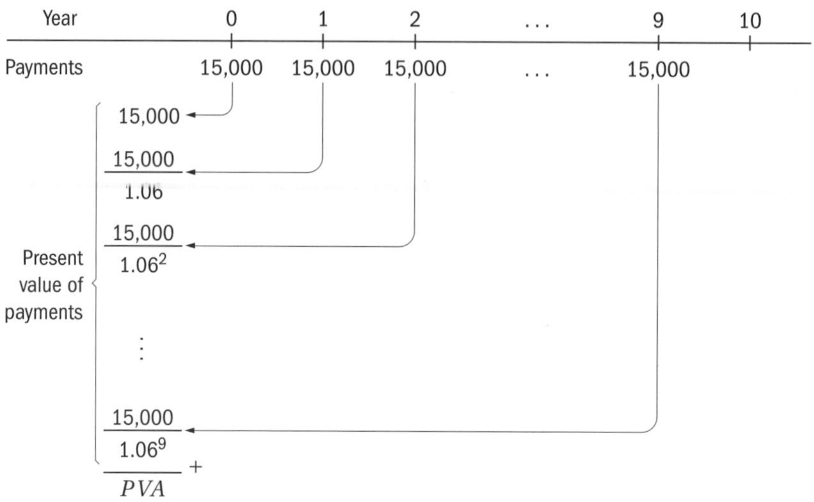

On this timeline we can see that from year 0 up to year 9 you get paid out €15,000 per year. The present value of the €15,000 which is paid out in year 0 equals €15,000. Hence in order to receive €15,000 now you should deposit €15,000 today, in year 0. The next year, in year 1, you would get paid out another €15,000 which today is worth $€ = \frac{15{,}000}{1.06}$, hence you should also deposit $€ = \frac{15{,}000}{1.06}$ today, in year 0. Continuing this argumentation we see that you should make a total deposit *PVA* of

$$PVA = 15{,}000 + \frac{15{,}000}{1.06} + \ldots + \frac{15{,}000}{(1.06)^9} \qquad (8.9)$$

in year 0. We recognize this expression as a geometric series with first term $a_1 = 15{,}000$, common ratio $r = \frac{1}{1.06} = (1.06)^{-1}$ and amount of terms $n = 10$. The result is

$$PVA = 15{,}000 \times \frac{\left((1.06)^{-1}\right)^{10} - 1}{(1.06)^{-1} - 1}$$

which is 1.06 times bigger than the *PVA* in equation (8.6). The result is:

$$\begin{aligned} PVA &= 1.06 \times 110{,}401.31 \\ &= 117{,}025.39 \end{aligned}$$

So, in order to get paid out $PMT = €15{,}000$ at the end of the year for the next 10 years, where the first payment occurs right now, you should deposit €117,025.39 today, in year 0. This deposit is the present value of this annuity, *PVA*.

This calculation can be generalized such that we can determine the present value of every annuity due.

THEOREM 8.7

Present Value of an Annuity Due

The present value PVA of an annuity due, which pays out PMT per year for the next n years, where the yearly interest rate equals i, is

$$PVA = PMT \times \frac{1-(1+i)^{-n}}{i} \times (1+i) \tag{8.10}$$

Compare this formula (8.10) with formula (8.3). Since the payments occur at the *beginning* of the year, this is an annuity *due*.

Explanation

This is exactly the same explanation as the one following Theorem 8.3: all the payments of an annuity due stay in the bank 1 year longer, so the final result must be bigger by a factor of $1 + i$.

THEOREM 8.8

Determining the Payments of an Annuity Due

The payments PMT of an annuity due can be determined by solving equation 8.10 for PMT.

8

8.3 Amortization

An amortization is a financial product whereby the amount of money borrowed from a bank is to be paid back in equal yearly payments. The calculations for annuities and amortizations are exactly the same, but the payments are going in the opposite direction: with an annuity you pay the bank an amount of money PVA which the bank pays back to you in equal yearly payments PMT, interest included. With an amortization the bank pays you an amount of money PVA which you pay back to the bank in equal yearly payments PMT, interest included. An amortization is of course an ordinary annuity, for it is not practical to borrow an amount of money from a bank and immediately pay back the first term. So if you want to borrow €400,000 against an interest rate of $i = 6\%$ and pay back this loan in $n = 30$ years then the yearly payments are the same €29,059.56 as we have found in Example 8.6.

Amortization with Nominal Interest

Students should notice that the payments of an annuity due are made at the beginning of each period. An example of annuity due could be the rent. Students usually pay their rent at the beginning of every month.

EXTEND YOUR KNOWLEDGE
What is the payback per year if you pay back forever? This is called a perpetuity. We let $n \rightarrow \infty$ in the equation

$$400{,}000 = PMT \times \frac{1-(1.06)^{-n}}{0.06}$$

For $n \rightarrow \infty$ the term $(1.06)^{-n}$ equals $\frac{1}{1.06} \times \frac{1}{1.06} \times \ldots$ and tends to 0. We thus have

$$400{,}000 = \frac{PMT}{0.06}$$

Which has the solution

$$PMT = 0.06 \times 400{,}000$$

Example 8.8 The Luxury Apartment

We shall explain the concept of an amortization with nominal interest where the periods between payments are not years, but months. Remind yourself of Theorem 6.6 before you proceed. Suppose that on the way to today's Mathematics class you saw a luxury apartment at the beach, costing only €400,000. Now you want to borrow this $PVA =$ €400,000 from a bank. The nominal interest charged is 6% per year, compounded monthly. And you would like to pay off this debt in, say, 30 years. How much will your monthly payback be? First of all we need to calculate the effective monthly interest, which is equal to $i = \frac{6\%}{12} = 0.5\%$, and the amount of months to pay back, this equals $n = 30 \times 12 = 360$. Now we can insert the information onto a time line, see figure 8.7.

8

We can insert these figures into formula (8.7) and solve the formula for PMT:

$$PVA = PMT \times \frac{1-(1+i)^{-n}}{i}$$

$$400{,}000 = PMT \times \underbrace{\frac{1-(1.005)^{-360}}{0.005}}_{166.791614}$$

Hence

$$PMT = \frac{400{,}000}{166.791614} = 2{,}398.20$$

Thus the monthly payments are all equal to €2,398.20.

FIGURE 8.7 Amortization with nominal interest

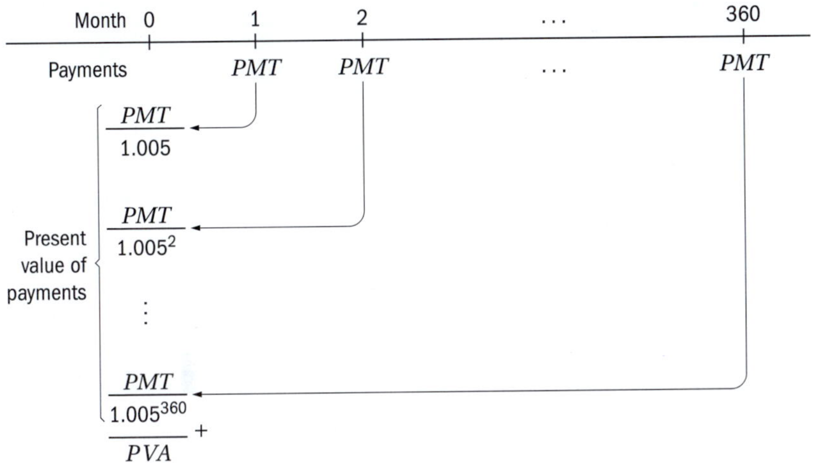

Finally, we can also calculate the amount of money which you can borrow if you know the monthly amount to be paid back, the number of instalment periods, and the monthly interest rate. We explain this using an example.

Example 8.9 Determining the Loan when the Instalments are Given

What amount of money can you borrow if you are able pay back $PMT = €800$ per month? We assume that nominal yearly interest is equal to 8% compounded monthly, so the effective interest is $i = \frac{8\%}{12} = 0.00667$, and you want to pay back your debt in, say, 20 years.
We recognize $n = 20 \times 12 = 240$ and $i = 0.00667$. Since the money is borrowed we are facing an ordinary annuity. We insert these numbers into formula (8.7):

$$\begin{aligned} PVA &= PMT \times \frac{1-(1+i)^{-n}}{i} \\ &= 800 \times \frac{1-(1.00667)^{-240}}{0.00667} \\ &= 95{,}614.97 \end{aligned}$$

So you can borrow €95,614.97 from the bank.

EXTEND YOUR KNOWLEDGE
Suppose you want to pay back €700 instead of €800 per month for the same amortization. How long does it take to pay back your debt? Just insert this value of PMT into the formula and solve for n:

$$95{,}614.97 = 700 \times \frac{1-(1.00667)^{-12n}}{0.00667}$$

Bringing all the terms to the left side, except $(1.00667)^{-12n}$ shows

$$(1.0067)^{-12n} = 0.088925929$$

Taking the logarithm and simplifying yields $n = 30.33499638$, a little less than 30 years, 4 months, 14 hours, 50 minutes, and 52 seconds (but you still have to pay back for 30 years and 5 months)

8.3.2 Interest Paid or Received

In Example 8.9, the amount of interest paid at the end of the term is equal to:

$$240 \times 800 - 95{,}614.97 = 96{,}385.03$$

If this were an annuity, the €96,385.03 would be the amount of interest you would receive. In general one can calculate the amount of interest paid or received using:

$$R = n \times PMT - PVA$$

Compare this formula with formula 7.2.

8.3.3 Remaining Debt

In this section we will analyze the remaining debt of an annuity. You can think of an annuity as either building up equity or paying off debt. Although the payments are all equal, equity doesn't build up at a constant rate: that's because at the beginning the debt is still high, so most of the payments are paying interest; toward the end, the remaining debt is small so very little of the payment goes toward interest (figure 8.8) (source: www.moneychimp.com).

8

Example 8.10 Remaining Debt of an Ordinary Annuity. The Luxury Apartment (Example 8.8) revisited

FIGURE 8.8 The remaining debt decreases exponentially with time

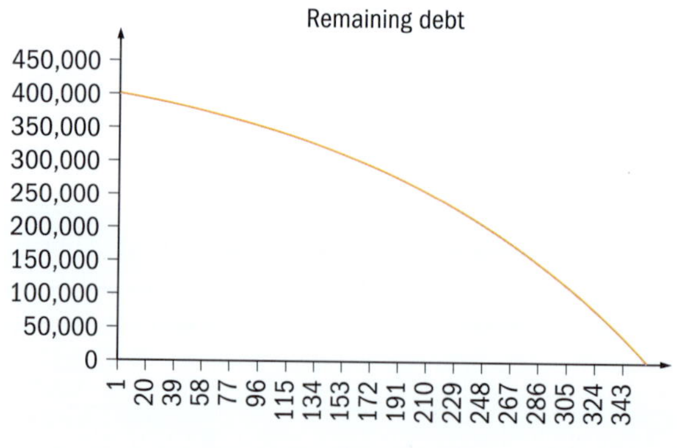

When 180 of the 360 payments are met, how much is the remaining debt? You might be tempted to say that this is $\frac{180}{360} = \frac{1}{2}$ of the original

debt, since half of the payments are met, but, as we shall see, this is not the case. Since we are dealing with a debt we must treat this as an ordinary annuity. Let us visualize our problem on a timeline first (see figure 8.9).

FIGURE 8.9 Remaining debt of an ordinary annuity

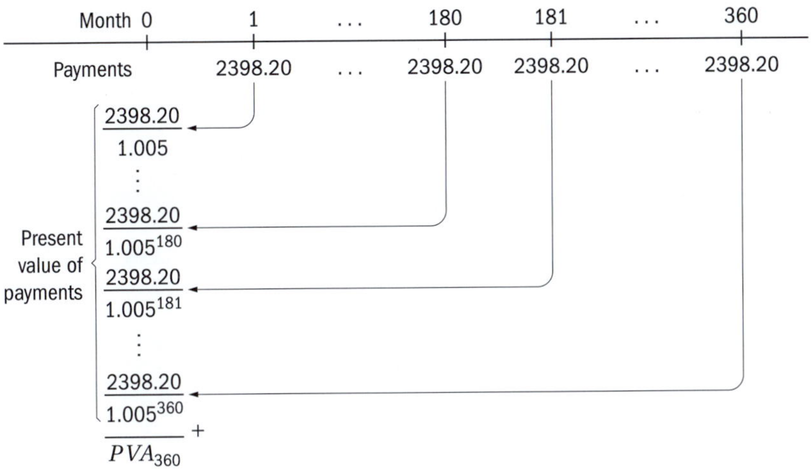

The part of the timeline just after the 180th payment looks like this:

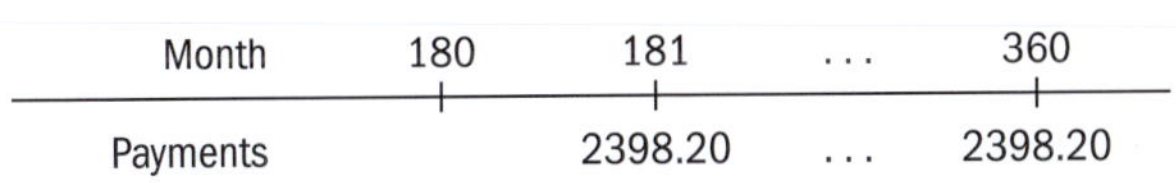

However, when we arrived at this point of time, month 180 is now, at present. Hence we can reset our stopwatch, resulting in replacing 180 by 0, and 181 by 1, etcetera, and 360 by $360 - 180 = 180$. This makes sense for there are still $360 - 180 = 180$ payments to be met. We can calculate all the present values of the payments at the new present and put these present values on a timeline:

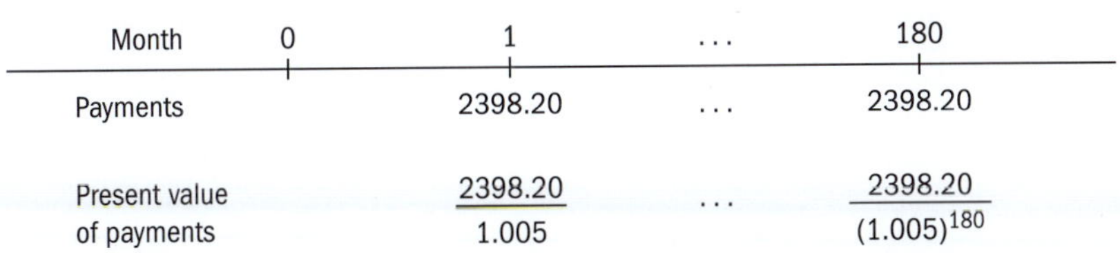

The present value of the amortization at month 180, PVA_{180}, is the remaining debt after 180 months. Inserting the figures of the timeline into formula (8.5), yields that the remaining debt after 180 months, PVA_{180}, equals:

$$
\begin{aligned}
PVA_{180} &= PMT \times \frac{1-(1+i)^{-n}}{i} \\
&= 2398.20 \times \frac{1-(1.005)^{-180}}{0.005} \\
&= 2398.20 \times 118.5035147 \\
&= 284{,}195.13
\end{aligned}
$$

As you can see, much less than half of the debt is paid off after half of the payments. This is because of the interest.

Analogous to Example 8.10 we can derive the following theorem:

THEOREM 8.9

Remaining Debt of an Ordinary Annuity

The remaining debt after k payments, PVA_k, is equal to

$$PVA_k = PMT \times \frac{1-(1+i)^{k-n}}{i} \qquad (8.11)$$

for an amortization which has to be paid off in n equal terms of PMT against an interest rate of i per term.

8

Summary

- Annuities deal with recurring payments. When there is only *one* payment, use the formulas (7.1) and (7.2).

- There are 4 types of annuities, which can be summarized in the following table:

Largest transaction \ Payments	Beginning	End
Present	*PVA* due	Ordinary *PVA*
Future	*FVA* due	Ordinary *FVA*

- When the payments occur at the *end* of the year (or any other time slot) we are dealing with an ordinary annuity.

- When the payments occur at the *beginning* of the year (or any other ordinary time slot) we are dealing with an annuity *due*.

- When dealing with time value of money, always draw a timeline first.

- The future value of an ordinary annuity equals

$$FVA = PMT \times \frac{(1+i)^n - 1}{i}$$

  when n deposits of PMT are made at the end of every year for n years, and the interest rate equals i.

- The future value of an annuity due equals

$$FVA = PMT \times \frac{(1+i)^n - 1}{i} \times (1+i)$$

  when n deposits of PMT are made at the beginning of every year for t years, and the interest rate equals i.

- Finding the payments PMT which should be made in order to have an amount of FVA in a savings account after n years against an interest rate of i can be found by solving either

$$FVA = PMT \times \frac{(1+i)^n - 1}{i}$$

  or

$$FVA = PMT \times \frac{(1+i)^n - 1}{i} \times (1+i)$$

  for PMT. One solves the first equation in case of an ordinary annuity, and the second in case of an annuity due.

- The total amount of interest received, R, equals

$$R = FVA - n \times PMT$$

  where FVA is the balance of a savings account after n payments of PMT against an interest rate of i.

- The present value of an ordinary annuity equals

$$PVA = PMT \times \frac{1-(1+i)^{-n}}{i}$$

  when n deposits of PMT are made at the end of every year for n years, and the interest rate equals i.

- An amortization is an ordinary annuity.

- The present value of an annuity due equals

$$PVA = PMT \times \frac{1-(1+i)^{-n}}{i} \times (1+i)$$

when n deposits of PMT are made at the beginning of every year for n years, and the interest rate equals i.

- Finding the payments PMT which are to be met, or to be received, from an annuity with a duration of n years against an interest rate of i can be found by solving either

$$PVA = PMT \times \frac{1-(1+i)^{-n}}{i}$$

or

$$PVA = PMT \times \frac{1-(1+i)^{-n}}{i} \times (1+i)$$

for PMT. One solves the first equation in case of an ordinary annuity, and the second in case of an annuity due.

- The total amount of interest paid or received, R, equals

$$R = n \times PMT - PVA$$

where PVA is the loan or the price of an annuity against an interest rate of i. The annuity has a duration of n years, and PMT are the payments.

- The remaining debt after k payments, PVA_k, is equal to

$$PVA_k = PMT \times \frac{1-(1+i)^{k-n}}{i}$$

for an amortization which has to be paid back in n equal terms of PMT against an interest rate of i per period.

Exercises

8.1 At the end of each year Angela invests €3,500 in a savings account that earns interest at 3% per year. Find the value of her investment after twelve deposits.

8.2 At the beginning of each month, Priyanka deposits €15 in a savings account that earns interest at a rate of 0.6% per month on the minimum monthly balance. How much are her investments worth after 4 years?

8.3 At the end of every month, Soufiane deposits €100 in a savings account that earns interest at the rate of 0.4% per month on the minimum monthly balance.

a How much is his investment worth at the end of the 5th year (that is 60 deposits)?
b How much interest did Soufiane receive?
c How much is the yearly interest rate?
d If Soufiane deposits his $12 \times$ €100 = €1,200 at the end of each year, how much money is in his savings account at the end of the 5th year if we assume the yearly interest rate equals the interest rate found in **c**? Why is this less?

8.4 Suppose you deposit an amount *PMT* in a savings account at the beginning of the year, starting at the end of this year, and ending at the end of year 10. The interest rate equals 3.6%.

a What deposits should be made in order to have €10,000 at the end of the 10th year from now?
b How much interest is received?

8.5 Suppose you deposit an amount *PMT* in a savings account at the end of each year, starting this year and ending at the end of year 7. The interest rate equals 4.7%.

a What deposits should be made in order to have €12,345 8 years from now?
b How much interest is received when the final target is met?

8.6 At the beginning of each year, Sara deposits €1,000 in her savings account. The annual interest rate is 5%. How long does it take before Sara has €10,000?

8.7* What interest rate i is needed if you want to have €3,200 in your savings account in 3 years from now, and the yearly deposits to be made at the end of each year are all equal to €1,000?

8.8* If at the beginning of each year a deposit of *PMT* is made in a savings account earning an interest rate of i, show how the balance at the beginning of the nth year is equal to

$$FVA = PMT \times \frac{(1+i)^{n+1} - 1}{i}$$

The first deposit is made at the beginning of year 0, and the last at the beginning of year n. Hence there are $n+1$ deposits.

8.9

a What is the value of €10 if the interest rate is $i=3\%$ after 1 year? And after 2 years?
b What is the present value of €10 next year if the interest rate is $i=3\%$? And the year after next?

8.10 What is the present value of annuity which pays out €1,500 per year in the following 25 years if the interest rate is 5%?

8.11

a What is the present value of an annuity which pays out €10,000 at the end of each year for the next 10 years if the interest rate is equal to 6%?
b What is the present value of an annuity which pays out €20,000 at the end of each year for the next 10 years if the interest rate is equal to 6%?
c What is the present value of an annuity which pays out €10,000 at the end of each year for the next 10 years if the interest rate is equal to 12%?

8.12

a What is the present value of an annuity which pays out €10,000 at the beginning of each year for the next 10 years if the interest rate is equal to 6%?
b What is the present value of an annuity which pays out €20,000 at the beginning of each year for the next 10 years if the interest rate is equal to 6%?
c What is the present value of an annuity which pays out €10,000 at the beginning of each year for the next 10 years if the interest rate is equal to 12%?

8

8.13

a How much is the payback per annum (year) for a loan of €56,502.23 which is to be paid back in the next 10 years against an interest rate of 12%?
b What are the similarities with **8.11c?**

8.14 How much is the payback per year for a loan of €100,000 which is to be paid back in the next 20 years against an interest rate of 5%?

8.15

a Suppose you have won the 27.5 million euro jackpot from the lottery. How much can you withdraw at the beginning of each year if you want to have received all your prize in 80 years. Assume the interest rate equals 3%.
b Suppose you want to withdraw €884,051.46 at the beginning of the year for the next 80 years, how much should you put in your bank account now if the interest rate is 3%?

8.16 As a simplified model for a retirement plan consider the following example: at the end of each year someone puts €1,200 in a savings account. She or he does that for 30 years, i.e. there are 30 payments; the first is made at the end of year 0, the last at the end of year 29. At the end of year 29 this person buys an annuity which costs the same as the balance in her or his savings account. Assume the term of this annuity is also 30 years, i.e. the first payment is at the end of year 30, and the last at the end of year 59. The question to be answered in this exercise is how much the yearly payments *PMT* during the retirement are. Assume the interest rate of the savings account and the annuity is equal to 3%.

a What would be the balance *K* in the savings account after the last deposit is made at the end of year 29?
b Now determine the yearly payments of the annuity which costs *K* as found in **a**.
c How much interest would she or he receive in total?

Please note that the payments are not corrected for inflation, hence the values of these future payments would be far less.

8.17 Suppose you want to save some money for your retirement in 25 years from now. You start saving at the end of this year, and the last deposit in your savings account will be at the end of the 24th year. You would like to get €10,000 per year starting at the end of year 25 and ending at the end of year 49. Assume the interest rate is 4%.

a What is the present value *PVA* of an annuity which pays out €10,000 per year for the next 25 years?
b How much should the yearly deposits be in order to have *PVA* in your savings account at the end of year 24, when you will buy this annuity?

8.18 In this exercise we will investigate whether it makes any difference to pay off a debt with equal yearly payments, or to put these payments in a savings account with the same interest and term, and pay off the debt in a single payment at the end of the term. Suppose you have a debt of, say, €20,000 which is to be paid back in 10 equal yearly payments. The interest rate is equal to 5%.

a Determine the payments *PMT* for this loan.
b How large does this debt of €20,000 become if nothing is be paid back at all over 10 years?
c If all the payments *PMT*, as found in **a**, are put in the savings account, what will the balance of this savings account be after 10 years?
d* Where does the difference between the answers of **b** and **c** come from?

8.19 Salvatore borrows €18,000 from the bank at a rate of 4.2% interest per annum. He repays it in ordinary instalments at the end of each month. The loan is to be paid off in 3 years.

a Calculate his monthly repayments.
b Calculate the remaining debt after 1 year (= 12 repayments).
c What would be the total interest paid on the loan after those 3 years?

8.20 Moty wants to buy a sports car for €45,000. He can pay 10% of this amount himself. He wants to borrow the rest from the ABM bank. The bank offers him a nominal rate of interest of 10.8% per year. He needs to repay the loan in ordinary instalments at the end of each month. The loan is to be paid off in 5 years.

a Calculate his monthly payment.
b Calculate the total interest paid on the loan after those 5 years.

8.21 Chin Chin decides to invest €25,000 of the money she got from selling her house in the Silver Lining Annuity Bank. The bank promises Chin Chin a sum *PMT* at the end of every year. The (yearly) rate the bank offers is 7.2% over a 15 year period. Calculate *PMT* for this situation.

8.22 How much does it cost to buy an annuity of €12,000 per year for 20 years if the rate of interest is 4.3% per annum?

8.23 Zainab borrows €10,000 from the bank at a 6% interest per annum. She repays it in ordinary installments at the end of each month. The loan is to be paid off in 5 years.

a Calculate her monthly repayments.
b What would be the total interest paid on the loan after those 5 years?
c Calculate the remaining debt after 2.5 years; is this half of the original debt?
d Calculate the new monthly payments if the interest rate after 2.5 years changes to 15%.

8.24* Suppose John borrows an extra €100 per month from the DUO in order to finance his studies. He receives the first €100 at the beginning of the first month of his studies, and the last €100 at the beginning of the last month of his studies. The duration of his studies is exactly 4 years, and the interest rate charged by the DUO is 0.124% per month.

a What is the total debt at the beginning of the first month after John has finished his studies?
b John can wait for 2 years before he has to start paying back his debt. How much debt does he have on the last day of the month when he starts paying back his debt? (Since most bills have to be paid at the end of the month, John actually has 25 months between finishing his studies and the first repayment.)
c John's debt has to be paid back in 15 years. How much are his monthly repayments?
d The minimum monthly payments to DUO however are €45. Over how many months does John have to pay back his debt?
e How much interest does John pay? (Hint: use the unrounded answers to **d** multiplied by €45.)

bank

9
Matrices and Markov Chains

At the end of this chapter you should be able to:
- understand the concept of a matrix
- determine the stationary state for a given transition matrix
- understand the operations involved in matrices (addition, subtraction, and multiplication)
- find the determinant of a matrix
- solve a system of linear equations using matrices (Cramer method)
- identify regular Markov chains

This chapter discusses the concept of matrix, which is very useful in business as it helps to represent information displayed in a table. Matrix also helps in solving systems of two or more linear equations. This chapter will also show the application of a matrix in the business world.

Input-output analysis

Input-output analysis is a basic method of quantitative economics that portrays macroeconomic activity as a system of interrelated goods and services. In particular, the technique observes various economic sectors as a series of inputs of source materials (or services) and outputs of finished or semi-finished goods (or services). The field is most identified with the work of Wassily Leontief (1906–1999), who was awarded the 1973 Nobel Prize in Economics for his pioneering work in the area. Leontief once explained input-output analysis as follows: 'When you make bread, you need eggs, flour, and milk. And if you want more bread, you must use more eggs. There are cooking recipes for all the industries in the≈economy.' And hence, one industry's output is another's input, and the chain continues.

Read more: Input-Output Analysis http://www.referenceforbusiness.com/encyclopedia/Inc-Int/Input-Output-Analysis.html#ixzz1sUyQWNmq

9.1 What is a Matrix

A matrix is a rectangular array of numbers, such as:

$$\begin{pmatrix} 10 & 7 \\ 12 & 6 \end{pmatrix}$$

A matrix is not only a convenient shorthand for describing information, but has also lots of computational advantages.
Imagine you have two different jobs, e.g. typing and babysitting. Suppose you earn €10 per hour by typing and €7 per hour by babysitting. How much are your earnings this week after 2 hours of typing and 3 hours of babysitting?

$$10 \times 2 + 7 \times 3 = €41$$

This can also be written in matrix multiplication form:

$$(10 \quad 7)\begin{pmatrix} 2 \\ 3 \end{pmatrix} = 10 \times 2 + 7 \times 3 = 41$$

Next week you want to spend 4 hours on typing and 1 hour on babysitting. What are your earnings for this week and next week?

$$(10 \quad 7)\begin{pmatrix} 2 & 4 \\ 3 & 1 \end{pmatrix} = (41 \quad 47)$$

This week you've earned €41, and next week you'll earn €47; in total this will be €88. The amount of hours you spend working per job in these weeks would be

$$\begin{pmatrix} 2 \\ 3 \end{pmatrix} + \begin{pmatrix} 4 \\ 1 \end{pmatrix} = \begin{pmatrix} 6 \\ 4 \end{pmatrix} \tag{9.1}$$

so you would earn in total

$$(10 \quad 7)\begin{pmatrix} 6 \\ 4 \end{pmatrix} = €88$$

as we already knew. In another city you can earn €12 per hour for typing and €6 per hour for babysitting. If you work the same hours, how much can you earn there?

$$(12 \quad 6)\begin{pmatrix} 6 & 4 \\ 3 & 1 \end{pmatrix} = (24 + 18 \quad 48 + 6) = (42 \quad 54)$$

For comparison we put the jobs from both cities and the hours of both weeks in 2 matrices and multiply them:

$$\begin{pmatrix} 10 & 7 \\ 12 & 6 \end{pmatrix}\begin{pmatrix} 2 & 4 \\ 3 & 1 \end{pmatrix} = \begin{pmatrix} 41 & 47 \\ 42 & 54 \end{pmatrix}$$

This example illustrated the concepts of matrices, their addition and multiplication, which can be defined formally.

9.2 Matrices and Operations on Matrices

Definition 9.1 Matrix
A matrix is a rectangular array of numbers.
A matrix consisting of n rows and m columns is called an $n \times m$ matrix. The entry on the $_i$th row and the $_j$th column is denoted by a_{ij}.
For example, the matrix

$$A = \begin{pmatrix} 5 & -3 & 1 \\ 2 & 6 & -4 \end{pmatrix}$$

is a 2×3 matrix, the entry a_{12} is equal to -3.

A memory aid for distinguishing between rows and columns is to remember the positioning of terraced houses.

Definition 9.2 Addition of Matrices
The addition of 2 matrices of an equal size is defined by the addition of the corresponding entries, i.e.

$$\begin{pmatrix} a_{11} & a_{12} \\ a_{21} & a_{22} \end{pmatrix} + \begin{pmatrix} b_{11} & b_{12} \\ b_{21} & b_{22} \end{pmatrix} = \begin{pmatrix} a_{11}+b_{11} & a_{12}+b_{12} \\ a_{21}+b_{21} & a_{22}+b_{22} \end{pmatrix} \qquad (9.2)$$

Example 9.1

$$\begin{pmatrix} 1 & 2 \\ 3 & 4 \end{pmatrix} + \begin{pmatrix} 2 & 4 \\ 6 & -8 \end{pmatrix} = \begin{pmatrix} 1+2 & 2+4 \\ 3+6 & 4-8 \end{pmatrix} = \begin{pmatrix} 3 & 6 \\ 9 & -4 \end{pmatrix}$$

Compare this addition with equation (9.2).
We now turn to the multiplication of matrices.

Definition 9.3 Multiplication of Matrices
The multiplication of 2×2 matrices is defined by:

$$\begin{pmatrix} a & b \\ c & d \end{pmatrix} \begin{pmatrix} e & f \\ g & h \end{pmatrix} = \begin{pmatrix} ae+bg & af+bh \\ ce+dg & cf+dh \end{pmatrix} \qquad (9.3)$$

Considering the following example makes it much easier to understand how to perform this multiplication. We calculate the product:

$$\begin{pmatrix} 1 & 2 \\ 3 & 4 \end{pmatrix} \begin{pmatrix} 10 & 9 \\ 8 & 7 \end{pmatrix}$$

We write this product in the following arrangement and then calculate the product.

$$\begin{matrix} & \begin{pmatrix} 10 & & 9 \\ 8 & & 7 \end{pmatrix} \\ \begin{pmatrix} 1 & 2 \\ 3 & 4 \end{pmatrix} & \begin{pmatrix} 1\times 10+2\times 8 & 1\times 9+2\times 7 \\ 3\times 10+4\times 8 & 3\times 9+4\times 7 \end{pmatrix} = \begin{pmatrix} 26 & 23 \\ 62 & 55 \end{pmatrix} \end{matrix}$$

Matrix multiplication is not commutative. That means that the order of multiplication of two matrices cannot be reversed, i.e.

$$AB \neq BA$$

This can be illustrated using the following example:

$$\begin{pmatrix} 1 & 1 \end{pmatrix} \begin{pmatrix} 1 \\ 1 \end{pmatrix} = 2$$

Whereas

$$\begin{pmatrix} 1 \\ 1 \end{pmatrix} \begin{pmatrix} 1 & 1 \end{pmatrix} = \begin{pmatrix} 1 & 1 \\ 1 & 1 \end{pmatrix}$$

Hence

$$\begin{pmatrix} 1 & 1 \end{pmatrix} \begin{pmatrix} 1 \\ 1 \end{pmatrix} \neq \begin{pmatrix} 1 \\ 1 \end{pmatrix} \begin{pmatrix} 1 & 1 \end{pmatrix}$$

Compare this with exercise 1.3 h and i.

Next to matrix multiplication as defined above, a matrix can be multiplied by a single number as well. This number is called a scalar. Consider the earnings matrix from the beginning of this section:

$$\begin{pmatrix} 10 & 7 \\ 12 & 6 \end{pmatrix}$$

If you want to know your wages in Yuan instead of euros, you have to multiply all the entries of the matrix by the conversion factor; this is the scalar. This scalar was as follows at the time of writing:

€1 = ¥8.83.

Hence the earnings matrix equals:

$$8.83 \begin{pmatrix} 10 & 7 \\ 12 & 6 \end{pmatrix} = \begin{pmatrix} 88.30 & 61.81 \\ 105.93 & 52.98 \end{pmatrix}$$

This example illustrates the definition of scalar multiplication of a matrix:

Definition 9.4 Scalar Multiplication.
Let s be a scalar, and A be any matrix. The scalar multiplication

$$sA$$

is defined by

$$s\begin{pmatrix} a_{11} & a_{12} \\ a_{21} & a_{22} \end{pmatrix} = \begin{pmatrix} sa_{11} & sa_{12} \\ sa_{21} & sa_{22} \end{pmatrix} \tag{9.4}$$

This just means that all the entries are multiplied by this scalar s.

9.3 Determinant and the Cramer's rule

Gabriel Cramer (figure 9.1) was a Swiss mathematician who introduced Cramer's rule in 1750 - a method for the solution of linear equations, which revived interest in the use of determinants; this was called Cramer's paradox, and formed the concept of utility in Mathematics.
Cramer was born and educated in Geneva, where at the age of 20 he became professor of Mathematics at the Académie de la Rive. In 1750 he was made professor of Philosophy. Cramer travelled throughout Europe and met leading Mathematicians. Cramer's paradox revolves around a theorem formulated by Scottish mathematician Colin Maclaurin, who stated that two different cubic curves intersect at nine points.

FIGURE 9.1 Gabriel Cramer (1704–1752)

Cramer pointed out that the definition of a cubic curve - a single curve - is that it is determined itself by nine points.
Cramer's concept of utility now provides a connection between the theory of probability and mathematical economics.

His major work is *Introduction à l'analyse des lignes courbes algébriques*, 1750, in which he set out Cramer's rule (source: www.cartage.org.lb).

Cramer's rule is a method involving the determinant of coefficients. The rule is mainly used for solving a system of linear equations. Cramer's rule for solving a system of two equations (with two variables) goes as follows. Given a system of equations:

$$\begin{cases} ax+by=e \\ cx+d=f \end{cases}$$

the system can be represented in a form of matrix:

$$\begin{pmatrix} a & b \\ c & d \end{pmatrix}\begin{pmatrix} x \\ y \end{pmatrix}=\begin{pmatrix} e \\ f \end{pmatrix}$$

Using Cramer's rule to solve the system we first need to calculate the determinant D, which is given by:

$$\begin{aligned} D &= \begin{vmatrix} a & b \\ c & d \end{vmatrix} \\ &= ad-bc \end{aligned}$$

Now, for $D \neq 0$ the solutions x and y are given by:

$$x = \frac{\begin{vmatrix} e & b \\ f & d \end{vmatrix}}{D} = \frac{ed-fb}{D}, \text{ and } y = \frac{\begin{vmatrix} a & e \\ c & f \end{vmatrix}}{D} = \frac{af-ce}{D}$$

9

Use Cramer's rule to solve this system of equations:

$$\begin{cases} 3x-7y=2 \\ 6x-13y=4 \end{cases}$$

The system can be represented in a form of matrix:

$$\begin{pmatrix} 3 & -7 \\ 6 & -13 \end{pmatrix}\begin{pmatrix} x \\ y \end{pmatrix}=\begin{pmatrix} 2 \\ 4 \end{pmatrix}$$

Using Cramer's rule to solve the system we first need to calculate the determinant D, which is given by:

$$\begin{aligned} D &= ad-bc \\ &= 3\times(-13)-6\times(-7) \end{aligned}$$

Now the solutions x and y are given by:

$$x=\frac{\begin{vmatrix} 2 & -7 \\ 4 & -13 \end{vmatrix}}{D} = \frac{2\times(-13)-4\times(-7)}{3}=\frac{2}{3}$$

$$y = \frac{\begin{vmatrix} 3 & 2 \\ 6 & 4 \end{vmatrix}}{D} = \frac{3 \times 4 - 6 \times 2}{3} = \frac{0}{3} = 0$$

So $x = \frac{2}{3}$ and $y = 0$

9.4 Markov Chains

A Markov analysis is a statistical technique used in forecasting the future behavior of a variable or system whose current state or behavior does not depend on its state or behavior at any time in the past: in other words, it is random. For example, in the flipping of a coin, the probability of a flip coming up heads is the same regardless of whether the previous result was heads or tails. In accounting, Markov analysis is used in estimating bad debt or uncollectible accounts receivable. In marketing, it is used in modeling future brand loyalty of consumers based on their current rate of purchases and repurchases. In quality control, Markov analysis is applicable to common-cause problems and other sequence-dependant events, and can handle system degradation. The Markov analysis was named after its inventor, the Russian mathematician and a probability theory pioneer Andrei Andreevich Markov (1856–1922) (figure 9.2) (source: www.businessdictionary.com).

FIGURE 9.2 Andrei Andrevich Markov (Андре́й Андре́евич Ма́рков) (1856–1922)

A Markov chain describes the transitions and distributions in percentages between different states. In this syllabus we explore the following model: Suppose there are only 2 firms (states), firm X and firm Y, and that the flow of customers from year to year is given by the flow chart which is shown in figure 9.3.

FIGURE 9.3 Flow chart of the flow of customers of firms X and Y

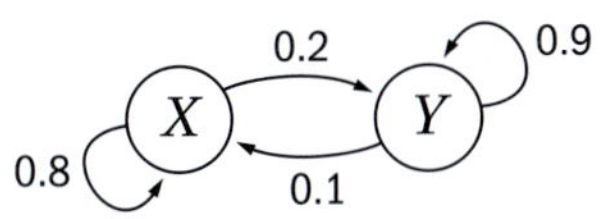

This states that from year to year 20% of the customers from firm X switches to firm Y, and the rest, 80% of the customers, stay with firm X. Furthermore, 10% of Y's customers switch to X and 90% stay. We can write that information into a table:

→	X	Y
X	0.8	0.2
Y	0.1	0.9

And in a matrix:

$$M = \begin{pmatrix} 0.8 & 0.2 \\ 0.1 & 0.9 \end{pmatrix} \tag{9.5}$$

This matrix M is called a transition matrix. Note that the entries of the flow chart, the table and the matrix are equal. Also note that the total of the entries per row add up to 1. This means that all the customers stay in this system of 2 firms, they either move to the other firm, or stay with their own. Now suppose that today, in year 0, $x_0 = 70\%$ of the customers are with x, and the rest, $y_0 = 30\%$, is with y. In a matrix:

$$(x_0 \quad y_0) = 0.7$$

How will the distribution (x_1, y_1) of customers be in the next year, year 1? The answer is given by the following matrix multiplication:

$$(x_1 \quad y_1) = (0.7 \quad 0.3)\begin{pmatrix} 0.8 & 0.2 \\ 0.1 & 0.9 \end{pmatrix} = (0.59 \quad 0.41)$$

Hence $x_1 = 59\%$ of the customers are with X and $y_1 = 41\%$ with Y. And the year after, year 2?

$$(x_2 \quad y_2) = (0.59 \quad 0.41)\begin{pmatrix} 0.8 & 0.2 \\ 0.1 & 0.9 \end{pmatrix} = (0.513 \quad 0.487)$$

After 2 years the distribution of customers over (X, Y) is $(x_2, y_2) = (51.3\%, 48.7\%)$. Hence the distribution of customers in the next year can be calculated by multiplying the present distribution with the transition matrix. Now note that the distribution of the second year, (x_2, y_2) was calculated by multiplying the distribution of the first year by the transition matrix. The distribution of the first year was, however, obtained by multiplying the initial distribution by the transition matrix. Now:

$$\begin{aligned}(x_2 \quad y_2) &= (0.59 \quad 0.41)\begin{pmatrix}0.8 & 0.2\\0.1 & 0.9\end{pmatrix}\\ &= (0.7 \quad 0.3)\begin{pmatrix}0.8 & 0.2\\0.1 & 0.9\end{pmatrix}\begin{pmatrix}0.8 & 0.2\\0.1 & 0.9\end{pmatrix}\\ &= (0.7 \quad 0.3)\begin{pmatrix}0.8 & 0.2\\0.1 & 0.9\end{pmatrix}^2\\ &= (0.7 \quad 0.3)\begin{pmatrix}0.66 & 0.34\\0.17 & 0.83\end{pmatrix}\\ &= (0.513 \quad 0.487)\end{aligned}$$

Hence the distribution after 2 transitions is the initial distribution multiplied by the square of the transition matrix. This idea can be generalized using the following theorem.

THEOREM 9.1

Distribution After *n* Transitions

Let the initial distribution of a system with 2 states be (x_0, y_0), and the transition matrix M be given by:

$$\mathrm{M} = \begin{pmatrix}m_{11} & m_{12}\\m_{21} & m_{22}\end{pmatrix}$$

9

Then the distribution of the system after n transitions, (x_n, y_n), is given by

$$(x_n \quad y_n) = (x_0 \quad y_0)\begin{pmatrix}m_{11} & m_{12}\\m_{21} & m_{22}\end{pmatrix}^n$$

where

$$\begin{pmatrix}m_{11} & m_{12}\\m_{21} & m_{22}\end{pmatrix}^n = \underbrace{\begin{pmatrix}m_{11} & m_{12}\\m_{21} & m_{22}\end{pmatrix}\cdot\begin{pmatrix}m_{11} & m_{12}\\m_{21} & m_{22}\end{pmatrix}\cdot\ldots\cdot\begin{pmatrix}m_{11} & m_{12}\\m_{21} & m_{22}\end{pmatrix}}_{n \text{ times}}$$

Long Run Distributions

We can continue calculating the distribution of customers for year 3, year 4, and so on. If we keep on calculating these distributions forever, will it come to a stable long-term distribution? If this were the case, then this distribution would not change anymore from year to year, because it is stable. In a formula:

$$(x \quad y)\begin{pmatrix}0.8 & 0.2\\0.1 & 0.9\end{pmatrix} = (x \quad y) \qquad (9.6)$$

where (x, y) denotes this (unknown) stable distribution. We proceed calculating this distribution. First we calculate:

$$(x \quad y)\begin{pmatrix} 0.8 & 0.2 \\ 0.1 & 0.9 \end{pmatrix} = (0.8x+0.1y \quad 0.2x+0.9y)$$

The result $(0.8x+0.8y, 0.2x+0.9y)$ must be equal to (x, y) because of equation (9.6). We therefore have this set of simultaneous equations:

$$\begin{cases} 0.8x+0.1y &= x \\ 0.2x+0.9y &= y \end{cases}$$

Subtracting $0.8x$ from both sides of the first equation and $0.9y$ from both sides of the second equation shows us that:

$$\begin{cases} 0.1y &= 0.2x \\ 0.2x &= 0.1y \end{cases} \tag{9.7}$$

THEOREM 9.2

Notes

Note that the coefficients 0.1 and 0.2 are the transitions from Y to X and from X to Y respectively, as can be found in the flow chart.
The 2 equations in the system (9.7) are equivalent, so we can discard either of them, and then simplify the other:

$$y=\frac{0.2}{0.1}x=2x \tag{9.8}$$

In order to calculate the long-run distribution, we remember that X and Y always serve 100% of the market, in a formula: $x+y=1$. Substitution of $2x$ for y from equation (9.8) into $x+y=1$ gives:

$$x+2x=1$$

Hence:

$$x=\frac{1}{3}=0.3333=33.3\%$$

So, in the long run X enjoys 33.3% of the market. This result can be obtained very easily by using Theorem 9.2. Take a look at the flow chart (figure 9.3).

Stable Distributions

We know from the preceding calculations that a stable distribution exists. A stable distribution means that the fraction of customers moving from Y to X $(0.1y)$ must be equal to the fraction moving from X to Y $(0.2x)$. Or, put in other words, that the flow out of X equals the flow into X. Thus:

$$0.1y=0.2x \tag{9.9}$$

9

The possible solutions $\tilde{y} = 0.2$ and $\tilde{x} = 0.1$ certainly satisfy this equation, but they add up to $0.1 + 0.2 = 0.3$, where they need to add up to 1. Therefore, let us divide our solutions $\tilde{x}$ and $\tilde{y}$ by this $0.1 + 0.2$, i.e.

$$x = \frac{0.1}{0.1 + 0.2} = \frac{1}{3} = 33.3\%$$
$$y = \frac{0.2}{0.1 + 0.2} = \frac{2}{3} = 66.7\%$$

FIGURE 9.4 The flow in equals the flow out, hence the level of the water remains stable

These solutions x and y add up to 1 and satisfy equation (9.6). The numbers 0.1 and 0.2 can be read from the flow chart and inserted into the formulae for x and y immediately. In general we have:

THEOREM 9.3

Long-term distribution

Let (X, Y) be a system with 2 states, X and Y. Let the transitions between X and Y be given by the transition matrix M:

$$M = \begin{pmatrix} 1-a & a \\ b & 1-b \end{pmatrix}$$

or the equivalent flow chart (figure 9.6).

FIGURE 9.5 Flow chart of transitions between X and Y

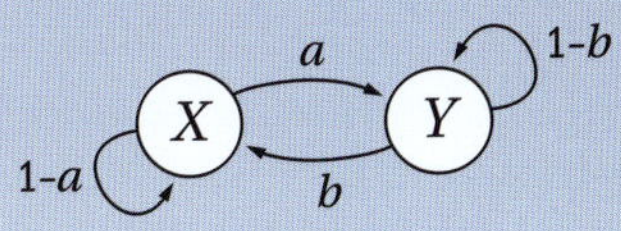

Then the long-term distributions of X and Y, x and y are given by:

$$x = \frac{b}{a+b}$$

$$y = \frac{a}{a+b}$$

9

Summary

- A matrix is a rectangular array of numbers, such as

$$\begin{pmatrix} 10 & 7 \\ 12 & 6 \end{pmatrix}$$

The entry a_{ij} is the number on row i and column j, i.e. $a_{12} = 7$.

- The addition of matrices of equal size is defined by the addition of the corresponding entries, i.e.

$$\begin{pmatrix} 10 & 7 \\ 12 & 6 \end{pmatrix} + \begin{pmatrix} 1 & 4 \\ 3 & 6 \end{pmatrix} = \begin{pmatrix} 10+1 & 7+4 \\ 12+3 & 6+6 \end{pmatrix}$$
$$= \begin{pmatrix} 11 & 11 \\ 15 & 12 \end{pmatrix}$$

- The multiplication of two 2×2 matrices is exemplified by

$$\begin{matrix} & \begin{pmatrix} 10 & 9 \\ 8 & 7 \end{pmatrix} \\ \begin{pmatrix} 1 & 2 \\ 3 & 4 \end{pmatrix} & \begin{pmatrix} 1 \times 10 + 2 \times 8 & 1 \times 9 + 2 \times 7 \\ 3 \times 10 + 4 \times 8 & 3 \times 9 + 4 \times 7 \end{pmatrix} \end{matrix}$$
$$= \begin{pmatrix} 26 & 23 \\ 62 & 55 \end{pmatrix}$$

Note that the order does matter, i.e. $AB \neq BA$.

$$M^n = \underbrace{M \cdot M \cdot \ldots \cdot M}_{n \text{ times}}$$

- A Markov chain describes the transitions and distributions in percentages between different states.

- If from year to year the transition of customers is given by, for example
  - 20% moves from X to Y
  - 10% moves from Y to X

we realize that the customers that don't move, will stay, and the flow chart can be drawn:

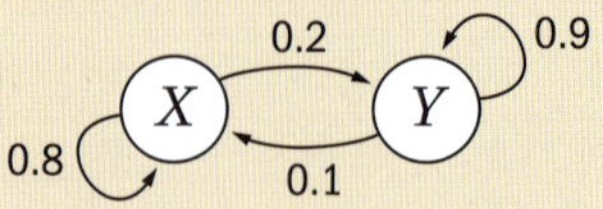

and we can determine the transition matrix:

$$\begin{pmatrix} 0.8 & 0.2 \\ 0.1 & 0.9 \end{pmatrix}$$

If the present distribution of the customers over the firms X and Y is given by, for example:

$$(x \quad y) = (0.7 \quad 0.3)$$

The distribution of the customers in the next year is given by:

$$(0.7 \quad 0.3)\begin{pmatrix} 0.8 & 0.2 \\ 0.1 & 0.9 \end{pmatrix} = (0.59 \quad 0.41)$$

And for the year after by:

$$(0.59 \quad 0.41)\begin{pmatrix} 0.8 & 0.2 \\ 0.1 & 0.9 \end{pmatrix} = (0.513 \quad 0.487)$$

- The long-term distribution can be determined from the flow chart or the transition matrix immediately by inserting the fractions of moving customers:

$$x = \frac{0.1}{0.1 + 0.2} = \frac{1}{3} = 33.3\%$$
$$y = \frac{0.2}{0.1 + 0.2} = \frac{2}{3} = 66.7\%$$

Exercises

9.1 Calculate:

a $\begin{pmatrix} 3 & 1 \\ 1 & 2 \end{pmatrix}\begin{pmatrix} 0 & 1 \\ 2 & 3 \end{pmatrix}$

b $\begin{pmatrix} 1 & 2 & 3 \\ 0 & -1 & 2 \end{pmatrix}\begin{pmatrix} 2 & -1 \\ 3 & 4 \\ 1 & 0 \end{pmatrix}$

c $\begin{pmatrix} 3 & 1 \\ 1 & 2 \end{pmatrix}\begin{pmatrix} 0 & 1 \\ 2 & 3 \end{pmatrix}+\begin{pmatrix} 1 & 2 & 3 \\ 0 & -1 & 2 \end{pmatrix}\begin{pmatrix} 2 & -1 & 3 \\ 4 & 1 & 0 \end{pmatrix}$

d $\begin{pmatrix} 1 & 0 & -1 \\ 0 & 1 & 2 \end{pmatrix}\left(\begin{pmatrix} 2 & 0 & \\ 1 & -10 & 1 \end{pmatrix}+3\begin{pmatrix} 0 & -1 \\ 1 & 2 \\ 0 & 0 \end{pmatrix}\right)+\begin{pmatrix} 2 & 1 \\ 3 & 0 \end{pmatrix}\begin{pmatrix} -1 & 2 \\ 0 & 1 \end{pmatrix}$

9.2 Calculate:

$\begin{pmatrix} 1 & 2 \\ 3 & 4 \end{pmatrix}+\begin{pmatrix} 0 & 1 \\ 2 & 3 \end{pmatrix}$ and $\begin{pmatrix} 0 & 1 \\ 2 & 3 \end{pmatrix}+\begin{pmatrix} 1 & 2 \\ 3 & 4 \end{pmatrix}$

Does the order of addition matter?

9.3 Calculate the matrix products

$\begin{pmatrix} 1 & 0 \\ 0 & 0 \end{pmatrix}\begin{pmatrix} 0 & 1 \\ 0 & 0 \end{pmatrix}$

and

$\begin{pmatrix} 0 & 1 \\ 0 & 0 \end{pmatrix}\begin{pmatrix} 1 & 0 \\ 0 & 0 \end{pmatrix}$

Does the order of multiplication matter?

9.4* Show that the addition of 2 × 2 matrices is commutative. For that calculate:
A + B and B + A

with $A=\begin{pmatrix} a_{11} & a_{12} \\ a_{21} & a_{22} \end{pmatrix}$ and $B=\begin{pmatrix} b_{11} & b_{12} \\ b_{21} & b_{22} \end{pmatrix}$

9.5 Find the determinant of the following matrices:

a $\begin{pmatrix} 1 & 1 \\ 1 & 1 \end{pmatrix}$

b $\begin{pmatrix} 1 & -1 \\ 1 & 1 \end{pmatrix}$

c $\begin{pmatrix} 2 & 4 \\ 6 & 8 \end{pmatrix}$

d $\begin{pmatrix} 3 & 5 \\ 1 & -2 \end{pmatrix}$

9.6 Use Cramer's Rule to solve the following systems of equations:

a $\begin{cases} x+y=3 \\ x-y=-1 \end{cases}$

b $\begin{cases} 2a+3b=13 \\ -a-3b=-14 \end{cases}$

c $\begin{cases} -2u+3v=2 \\ 4u-3v=-3 \end{cases}$

d $\begin{cases} 2x+3v=1 \\ 3x-2y=2 \end{cases}$

9.7 Sonix Company manufactures sofas and office chairs. During the year 2011, it manufactured a total of 50,000 sofas and office chairs. If the number of the office chairs produced were 10,000 more than the number of sofas, find the number of each product manufactured during year 2011.

9.8 Susan earns €15 per hour by tutoring, €10 per hour by typing, and €5 per hour by babysitting. The hours she worked at each of her jobs over a 4-week period are given by the following table:

	Week 1	**Week 2**	**Week 3**	**Week 4**
Tutoring	15	10	16	12
Typing	6	4	2	3
Babysitting	2	7	0	4

a Determine the earning matrix E and the matrix of hours worked per week H.

b Determine the matrix EH and give an interpretation of its elements (entries).

9.9 Calculate

$\begin{pmatrix} 1 & 1 \\ 0 & 1 \end{pmatrix}^1, \begin{pmatrix} 1 & 1 \\ 0 & 1 \end{pmatrix}^2$, and $\begin{pmatrix} 1 & 1 \\ 0 & 1 \end{pmatrix}^3$

9.10 In general there are two ways to get home after a day of work, by train and by car. At present 70% of the commuters travel by car, and the rest by train. A survey analysis over the past few years indicates that from one year to another, 10% of car commuters switch to travelling by train, and 15% of train commuters switch to travelling by car. Assume that this trend continues.

a Give the flow chart and the transition matrix for this Markov process.
b What percentage of commuters will travel by train in the next year?
c And, using your answer for **b**, what percentage of the market will travel by train in 2 years from now?
d What percentage of commuters will travel by train in the long run?

9.11 At the beginning of this academic year there was a smartboard eraser in every classroom of the school. Now, at the end of November, every teacher has to carry his or her own eraser around the building because all the erasers have disappeared mysteriously. This can bc illustrated by the following table of daily transitions:

	School	**Mysterious place**
School	0.95	0.05
Mysterious place	0	1

a Give the flow chart and the transition matrix for this Markov process.
b What percentage of the erasers will be in school 2 days after the opening of the academic year? Assume that all of the erasers were in this building at the beginning of this year.
c What percentage of the erasers is in the classrooms today? Either calculate the distribution of the erasers or approximate this distribution by the long-term distribution.

9

Solutions

Chapter 1

1.1 €700

1.2 €900

1.3
a True
b False
c True
d True
e True
f False
g True
h True
i False

1.4
a $\frac{2}{3}$
b 1
c $\frac{1}{2}$
d $\frac{1}{4}$
e $\frac{5}{6}$
f $-\frac{1}{6}$
g $\frac{11}{6} = 1\frac{5}{6}$
h $-\frac{24}{35}$

1.5
a $\frac{1}{4}$
b $\frac{2}{3}$
c $\frac{2}{3}$
d $-\frac{1}{3}$

e $-\frac{24}{35}$

f $-\frac{35}{24}$

g $\frac{3}{2} = 1\frac{1}{2}$

h 1

1.6 **a** 7
b 4
c 13
d 12
e 21
f 9
g 75
h 10

1.7 **a** $\frac{35}{32}$

b $\frac{165}{56}$

c $\frac{153}{70}$

d $\frac{-229}{45}$

e $\frac{-413}{120}$

f $\frac{1}{6}$

g $\frac{3}{2}$

h $\frac{3}{8}$

i 2

j $\frac{1}{2}$

k $\frac{3}{2}$

l $\frac{2}{3}$

1.8 **a** −4
b 2
c 10
d 14
e 13
f −3
g 8
h 2

1.9
- **a** 3
- **b** $3\sqrt{3}$
- **c** 6
- **d** 9
- **e** $\frac{3}{4}$
- **f** $\frac{3\sqrt{3}}{2\sqrt{2}}$
- **g** 3
- **h** 3

1.10
- **a** 14
- **b** -14.2
- **c** 12
- **d** -4
- **e** $\frac{3}{2}$
- **f** $\frac{1}{6}$
- **g** $\frac{5}{4}=1\frac{1}{4}$
- **h** $\frac{3}{7}$
- **i** $5\frac{2}{3}$
- **j** $4\frac{1}{6}$

1.11 The black one, because BEDMAS tells us to do division and multiplication from left to right.

1.12
- **a** 1
- **b** $\sqrt{6}$
- **c** 1
- **d** $\sqrt{2}$
- **e** 3
- **f** 7
- **g** 5
- **h** 7

1.13
- **a** C
- **b** $A=(-4,3)$, $B=(-2,-2)$, $C=(3,-4)$, $D=(1,-1)$, $E=(3,4)$
- **c** D

1.15
- **a** 0
- **b** $u+uv+v$

1.16
- **a** $23x+6y$
- **b** $7p+13q$
- **c** $54+14x$
- **d** $10-7x$

1.17
a $x^2+3x-28$
b x^2-y^2
c $a-b$
d $x^2+2xy+y^2$
e $2b^2-8b-42$
f $5x+3y$
g $10x^2-3x-7$

1.18 Hint: square both sides of the equation.

1.19
a $3(x+1)$
b $(x+1)(x+12)$
c $(x+1)(4x+1)$
d $(x+1)(4x+4)$

Chapter 2

2.1 $5, x+3=8$

2.2
a $6c+14a=20$ (with c and a amount of children's tickets and adult's tickets sold respectively)
b 35

2.3
a $x=2$
b $x=\frac{1}{2}$
c $x=-4$

2.4 Because the equations $x+2y=2$ and $y=-\frac{1}{2}x+1$ are equivalent.

2.5 **c** $x=0, y=1$

2.6 **c** $x=1, y=1$

2.7
a $x=1, y=2$
b $a=-1, b=5$
c $u=-\frac{1}{2}, v=\frac{1}{3}$
d $x=\frac{4}{5}, y=-\frac{1}{5}$

2.8
a $x=y=z=1$
b $a=1, b=2, c=3$
c $u=\frac{1}{2}, v=\frac{1}{3}, w=\frac{1}{4}$
d $x=\frac{4}{5}, y=-\frac{1}{5}, z=1$
e $u=\frac{3}{2}, v=\frac{1}{2}, w=-\frac{1}{2}$

2.9
- **a** $y=3x+1$, slope $=3$, y intercept $=1$
- **b** $y=-x-1$, slope $=-1-1$, y intercept $=-1$
- **c** $y=1$, slope $=0$, y intercept $=1$

2.10 1 = demand, 2 = supply, 3 = equilibrium, 4 = quantity, 5 = price

2.11 $P=1.80$ and $Q_d=Q_x=2{,}000$

2.12 Price = €1.73, quantity = 127

Chapter 3

3.1
- **a** $4x(x+1)+y^2$
- **b** $(y-5)^2$
- **c** $(6-2u)(6+2u)$
- **d** $(2x-y)(2x+y)$
- **e** $(3x-1)(3x+1)$
- **f** $3(x+1)^2$

3.2
- **a** $x=0$ and $x=1$
- **b** $x=0$ and $x=1$
- **c** $x=1$ and $x=2$
- **d** 13.52 and −0.52
- **e** 10.32 and 0.68
- **f** same solution as e

3.3
- **a** $x=-1$ and $x=3$
- **b** 10.32 and 0.68
- **c** $x=2$ and $x=3$
- **d** $x=3$ and $x=4$

3.4
- **a** $x=-5$
- **b** $x=3$
- **c** $x=4$
- **d** $x=-2, x=1$
- **e** $x=-1, x=2$
- **f** $x=-3, x=2$

3.6 Hint: since the quadratic is symmetric the top is in the middle of the x intercepts.

3.7
$$(2\times3)^4=\underbrace{(2\times3)\times(2\times3)\times(2\times3)\times(2\times3)}_{4\ times}=\underbrace{2\times2\times2\times2}_{4\ times}\times\underbrace{3\times3\times3\times3}_{4\ times}=2^4\times3^4$$

3.8
$$\frac{2^3}{2^5}=\frac{2\times2\times2}{2\times2\times2\times2\times2}=\frac{\not2\times\not2\times\not2}{2\times2\times\not2\times\not2\times\not2}=\frac{1}{2\times2}=\frac{1}{4}$$

3.9
- **a** 2
- **b** 2
- **c** 1
- **d** 8

e $\frac{1}{5}$

f 9

g 17

3.10 **a** x^3-y^3

b x^4-y^4

c x^3+x^2+x+1

d $3x^3-5x^2+7x-11$

3.11 **b** $x=0$, because $a^x=1$ for all $a\neq 1$. For if x increases by 1, y is multiplied by 1.

3.12 **a** $x=2$

b $x=1.262$

c $x=0.585$

d $x=0$

e $x=0.834$

f $x=1.357$

g $x=0.898$

h $x=-2.322$

3.13 **a** $x=-2$

b $x=1.892$

c $x=-3.265$

d $x=2.060$

e $x=-2$

f $x=0.631$

g $x=0$

h $x=-9$

3.14 **a** 2

b Hint: use $\frac{1}{x}=y$ implies $xy=1$

3.15 23 months

Chapter 4

4.1 **a** 0

b 3

c 3

d 2

e $15x^2$

f $6x$

g $6x^2$

h $12x^3$

i $0.5x^{-0.5}=\frac{1}{2\sqrt{x}}$

j $\frac{1}{2\sqrt{x}}$

k $\frac{-1}{x^2}=-\frac{1}{x^2}$

l $-\dfrac{1}{x^2}$

m $-2x^{-3} = \dfrac{-2}{x^3}$

n $\dfrac{-2}{x^3}$

o $-\dfrac{20}{x^5}$

p $-20x^{-5}$

4.2

a 3

b $6x$

c $e^x + 2$

d $2 + 2x$

e $3x^2 + 2x + 1$

f $e^x + \dfrac{1}{x}$

g $6x^2 + \dfrac{5}{x}$

h $6x + 6x^2$

i $\dfrac{1}{x} - e^x$

4.3

a $2x$

b $\ln x + 1$

c $2x \ln x + x$

d $2x$

e $\dfrac{2 + \ln x}{2\sqrt{x}}$

f 1

g $\dfrac{2 \ln x}{x}$

h 1

i $2 + 3\sqrt{x} + 2 \ln x$

4.4

a $\dfrac{1}{(x+1)^2}$

b $\dfrac{1}{(x+2)^2}$

c $\dfrac{-1}{(x^2-1)^2}$

d $\dfrac{1 - \ln x}{x^2}$

e $\dfrac{2 + \ln x}{2\sqrt{x}}$

f $\dfrac{1-x}{2\sqrt{x}(1+x)^2}$

4.5

a $3(x+1)^2$

b $\dfrac{1}{2\sqrt{x+1}}$

c $\dfrac{3\sqrt{x}}{\sqrt{2}}$

d $\dfrac{1}{2x}$

e $\dfrac{1}{2x}$

f $-2xe^{-x^2}$

g $\dfrac{2x+1}{x^2+x+1}$

h $\dfrac{2x}{(x^2-1)^2}$

i $3e^{3x}$

4.6
a $x=-1$
b $x=1$
c $x=0$
d $x=0$
e $x=1$
f $x=1$
g $x=-\dfrac{1}{2}$
h $x=0$
i $x=0$

4.7 $\dfrac{-b}{2a}$

4.8 Hint: $(f(x)+g(x)) = \dfrac{f(x+\Delta x)+g(x+\Delta x)-f(x)-g(x)}{\Delta x}$ as $\Delta x \rightarrow 0$

4.9
a Hint: $x \ln a = \ln a^x$
b $\ln a \times a^x$

Chapter 5

5.1
a 900
b 1,400
c €92, €95, €97

5.2
a €36,400, 364 chairs, 364 stools
b €38,500, 154 chairs, 616 stools

5.3 8,800

5.4 200

5.5 50,000

5.6
a 296
b 295.91
c Hint: $\Delta q = 301 - 300$

5.7
a $q_d = 900 - 500p$
b 0.9
c 450
d 5

5.8
a $50q - 200 - q^{1.5}$
b 710.6 k€
c 9

5.9 1000

5.10
a 4.76%
b 14.28% increase

5.11
a 1
b $\frac{1}{2}$

5.12
b 40, 20

5.13 10

5.14
c 3, 4

5.15
b €55, €55
c €11, €5.50
d €155
e €70.50

Chapter 6

6.1
a 10
b 12
c 55
d 30
e $x_1 + x_2 + x_3 x_3 + x_4 + x_5 + x_6 + x_7$
f $x_1 + x_2 + x_3 + x_4$
g $a_1 + a_2 + a_3$
h $x_1 + x_2 + x_3 + x_4 + x_5$
i $x_3 + x + 4 + x_5 + x_6 + x_7$
j 275
k 8
l 15

6.2
a 30
b 23
c 82
d 5
e −40
f −3
g $3a_k$
h 4
i 0

j 15.87
k 9.6

6.3 Different answers are possible.

a $\sum_{i=1}^{10} i$

b $\sum_{i=3}^{5} i$

c $\sum_{i=1}^{5} 3i$

d $\sum_{i=1}^{5} (9+3i)$

e $\sum_{i=2}^{3} i$

f $\sum_{i=0}^{5} 2^i$

g $\sum_{i=1}^{5} (3i-1)$

h $\sum_{i=0}^{5} (-3)^i$

6.4 Different answers are possible.

a $\sum_{i=1}^{10} x_i$

b $\sum_{i=3}^{5} x_i$

c $\sum_{i=1}^{4} x_i^2$

d $\sum_{i=2}^{4} (x_i^2-1)$

6.5
a 1
b 5
c 2.5
d 2.5
e 7.29
f 24.056
g 5.185
h 12380.432
i 40.651

6.6
a 110
b 555
c 77.5
d 7.5

e 58.59
f 269.056
g 178.945
h 12,382.9
i 307.68

6.7 50% increase

6.8 15.5% decrease

6.9 €971.09

6.10 €8,723.27, 24.62%

6.11 €6,141.25

6.12 3.55% yearly

6.13 6.06%

6.14 6.58 years ≈ 6 years and 7 months

6.15
a €123.10
b €150.90
c Interest received is €23.10 and €50.90 resp.

6.16
a €81.23
b €66.27, interest received is €18.77 and €33.73 resp.

6.17
a €14,150.94
b €13,349.95
c €8,375.92

6.18 23.45 years ≈ 23 years and 5 months.

6.19
a 3% for 20 years
b Hint: use $(1+i)^2 = 1 + 2i + i^2 > 1 + 2i$

6.20 €587,731.23

6.21 3.71%

6.22
a 20%
b 21%
c 21.9%
d 22.1%
e 22.1%
f 22.1%

6.23
a €1.20
b €1.21
c €1.22
d €1.22

e €1.22
f €1.22

6.24
a €2.49
b €2.59
c €2.70
d €2.71
e €2.72
f €2.72

6.25
a €19,906.56
b €20,749.94
c 21,567.76
d 21,704.58
e 21,740.30
f 21,646.25

6.26
a €3,563.06
b €3,568.33
c €3,572.83
d €3,573.53
e €3,573.71
f 3,573.74

6.27 50%

6.28 €491.59

6.29 €133,090.65

Chapter 7

7.1
a 55
b 55
c 110
d 100
e 42
f 0

7.2
a 4949
b 5050
c 5151
d 5252

7.3
a 129140162
b $\frac{1023}{1024}$
c 786429
d 3906

7.4
a 707
b 728
c 29,768

7.5
a 5050
b 250
c $30+40k$
d 97,625
e same as b
f $\frac{2047}{1024}$

7.6 7

7.7
a geometric; common ratio equals 1.03
b $a_1=500$, $r=1.03$, $n=4$
c 2091.81

7.8
a $a_1=500\times 1.03^3$, $r=\frac{1}{1.03}$, $n=4$
b 2091.81
c same series, only in reversed order

7.9 $2^{64}-1=18, 446, 744, 073, 709, 551, 615$ grains, which is about 50 billion tons, 73 times the present production per year.

Chapter 8

8.1 €49,672.10

8.2 €831.53

8.3
a €6,766.02
b €766.02
c 4.91%
d €6618.85, between the yearly deposits there is less money in the bank than in b.

8.4
a €819.00
b €1,810.00

8.5
a €1,530.11
b €1,634.23

8.6 $n\approx 7.98$. Since the interest is added to her bank account at the end of the year, she has to wait for 8 years.

8.7 6.52%

8.8 Calculate using a time line, find and then simplify

$$FVA=PMT\frac{(1+i)^n-1}{i}+PMT(1+i)^n$$

8.9
a €10.30, €10.61
b €9.71, €9.43

8.10 €21,140.92

8.11
a €73,600.87
b 2 × €73,600,87 = €147,201.74
c €56,502.23

8.12
a €78,016.92
b 2 × €78,016.92 = €156,033.84
c €63,282.50

8.13 €10,000

8.14 €8,024.26

8.15
a €884,051.46
b €27,500,000

8.16
a €57,090.50
b €2,912.72
c €51,381.60

8.17
a 156,220.80
b 3,751.17

8.18
a 2,590.09
b 32,577.89
c 32,577.87
d The answers of b and c were both rounded to the nearest cent.

8.19
a €533.03
b €12,249.79
c €1,189.08

8.20
a €876.53
b €12,091.80

8.21 €2,779.64

8.22 €158,835.95

8.23
a €193.33
b €1,599.80
c €5,373.31
d €215.89

8.24
a €4,948.70
b €5,104.41
c €20.39
d 123 months
e €704.47

Chapter 9

9.1 **a** $\begin{pmatrix} 2 & 6 \\ 4 & 5 \end{pmatrix}$

b $\begin{pmatrix} 11 & 7 \\ -1 & -4 \end{pmatrix}$

c $\begin{pmatrix} 13 & 13 \\ 3 & 1 \end{pmatrix}$

d $\begin{pmatrix} 0 & 1 \\ 1 & 13 \end{pmatrix}$

9.2 $\begin{pmatrix} 1 & 3 \\ 5 & 7 \end{pmatrix}$ and $\begin{pmatrix} 1 & 3 \\ 5 & 7 \end{pmatrix}$. No

9.3 $\begin{pmatrix} 0 & 1 \\ 0 & 0 \end{pmatrix}$ and $\begin{pmatrix} 0 & 0 \\ 0 & 0 \end{pmatrix}$. Yes

9.4 Use that $a_{ij} + b_{ij} = b_{ij} + a_{ij}$.

9.5 **a** 0
b 2
c −8
d −11

9.6 **a** $x = 1, y - 2$
b $a = 1, b = 5$
c $u = -\frac{1}{2}, v = \frac{1}{3}$
d $x = \frac{4}{5}, y = \frac{1}{5}$

9.7 30,000 chairs and 20,000 sofas

9.8 **a** $(15 \quad 10 \quad 5)\begin{pmatrix} 15 & 10 & 16 & 12 \\ 6 & 4 & 2 & 3 \\ 2 & 7 & 0 & 4 \end{pmatrix}$

b (295 225 260 230), €295 in week 1, €225 in week 2, €260 in week 3 and €230 in week 4.

9.9 $\begin{pmatrix} 1 & 1 \\ 0 & 1 \end{pmatrix}, \begin{pmatrix} 1 & 2 \\ 0 & 1 \end{pmatrix}$, and $\begin{pmatrix} 1 & 3 \\ 0 & 1 \end{pmatrix}$.

9.10 **a** 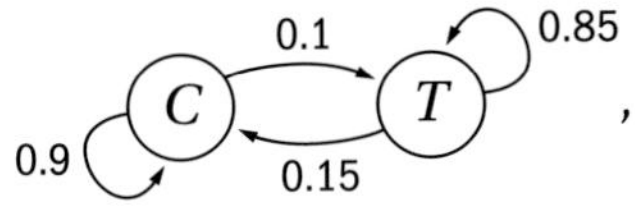

$, \begin{pmatrix} 0.9 & 0.1 \\ 0.15 & 0.85 \end{pmatrix}$

b 32.5%
c 34.4%
d 40%

9.11 **a** 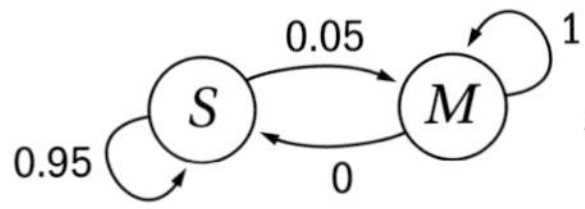

$, \begin{pmatrix} 0.95 & 0.5 \\ 0 & 1 \end{pmatrix}$

b 90.3%
c Direct calculation; after $13 \times 7 = 91$ days 0.93%. Approximation; 0%.

Bibliography

Craats, J. van de, Bosch, R. *Basisboek Wiskunde* (in Dutch), Pearson Education (2009), ISBN 978-90-430-1673-5.

Capinski, M., Zastawniak, T. *Mathematics for Finance*, Springer (2003), ISBN 1-85233-330-8.

Dondjio, I. *Syllabus Mathematics for IBMS Students* (2009), The Hague University.

Franck, R. *The economic naturalist*, Basic Books (2007), ISBN 0-465-00217-7

Jacques, I. *Mathematics for Economics and Business*, Prentice Hall (2009), ISBN 0-273-74329-5.

Slavin, S. *Business Math*, John Wiley & Sons (2007), ISBN 0-470-00719-8

Sloman, J., Wride, A. *Economics*, Pearson Education (2009), ISBN 0-273-71562-7.

Thomas, R.L *Using Mathematics in Economics* (1999), Pearson Education Ltd, ISBN 0-201-36050-0.

Zima, P., Brown, R.L. *Mathematics of Finance*, McGraw-Hill (1996), ISBN 0-07-008203-0.

Picture credits

ANP – Science Photo Library	Figure 9.2
Corbis	Chapter 5 - case
Imageselect	Figure 1.6, Chapter 2 – case, Chapter 4 – opening, Chapter 4 – case,
iStock	Chapter 1 – case, Figure 4.2
Jan van de Craats, faculteit FNWI, UvA	Figure 0.1
John Kuczala	Chapter 8 - case
Patrick Hermans	Chapter 9 - case
Reuters – Kimberly White	4.5 - case
Sloman, 2009	Figures 2.4 and 2.5
Spencer Platt – Getty Images	Chapter 6 - case
Wikipedia	Figure 1.1, Chapter 3 – case, Chapter 7 – case, Figure 9.1
Wouter Krasser and Irénée Dondjio	About the authors
Wouter Krasser	Chapter 1 – opening, Figures 1.3 and 1.4, Chapter 2 – opening, Figures 2.1, 2.2, 2.3, 2.6, 2.7 and 2.8, Chapter 3 – opening, Figures 3.1, 3.3, 3.4 and 3.5, Figures 4.3, 4.4, 4.5 and 4.6, Chapter 5 – opening, Figures 5.1, 5.2 and 5.3, Chapter 6 – opening, Chapter 7 – opening, 7.2 - running, Chapter 8 – opening, 8.2 – explanation, Figures 8.5 and 8.8, Chapter 9 – opening, Figures 9.4 and 9.5

Glossary

+	Addition
–	Subtraction
÷	Division
×	multiplication sign
x	x, a variable which' value is unknown
xy	$x \times y$
$x < y$	x is smaller than y, e.g. $1 < 2$
$x > y$	x is greater than y
$x \leq y$	x is smaller than, or equal to y
$x \geq y$	x is greater than, or equal to y
$x \neq y$	x is not equal to y
$\lvert x \rvert$	The absolute value of x
e	2.718381828459
Δ	Delta, difference
Σ	Summation sign
%	Percentage
Annum	year
j_m	interest rate of j compounded m times a year
FVIF	Future Value Interest Factor
$S_{n\angle i}$	Same as *FVIF*
PVIF	Present Value Interest Factor

$A_{n\angle i}$	Same as *PVIF*
FVIFA	Future Value Interest Factor Annuity
$S_{n\angle i}$	Same as *FVIFA*
PVIFA	Present Value Interest Factor Annuity
$a_{n\angle i}$	Same as *PVIFA*

Index

About the authors

Irénée Dondjio was born in Cameroon where he graduated in science specializing in Mathematics, Physics, Chemistry, Biology, Economics and Philosophy. After moving to the Netherlands in 1998 he changed his major to Business Studies, in order to combine his studies of science and business. He also speaks several European languages (French, English, German, Italian, and Dutch). He kept his fascination for mathematics while studying Business by organizing regular Mathematics support classes to help students who were less dedicated to the subject.

He graduated with an MBA in 2004 (at The Hague University) and has worked with several international companies such as Fujitsu and Western Union, but lecturing and consulting has always been a source of fascination for him. In 2006, he decided to start a lecturing career in Marketing, Economics and Mathematics with the ultimate goal of helping students to appreciate the benefit of mathematical knowledge in their daily lives, sparking his research to deepen the knowledge and understanding of Mathematics.

Irénée Dondjio traveled extensively throughout Australia, Asia, Europe and Africa, participating in lecturing and acting as a guest speaker at prominent universities. Irénée Dondjio is currently a guest lecturer at the Korean university of Hanyang where he lectures Marketing each year. Irénée Dondjio is currently pursuing his PhD in consumer behavior on the emerging African market, and course completion is expected by 2015. He currently works as a teacher at The Hague University.

After a career as a sailor, **Wouter Krasser** studied Stochastics and Financial Mathematics at the University of Amsterdam. As a postgraduate he lectured in several topics of Mathematics and Statistics at the Science Faculty of the University of Amsterdam. At present he is enjoying his job as a lecturer of Mathematics and Statistics at The Hague University of Applied Sciences. One of the aspects of Mathematics that is very attractive to him, is the fact that a problem can seem to be very difficult, until the often elegant and simple solution becomes clear. By writing this textbook he aims to make financial Mathematics more comprehensible for Business students and to introduce to them the mathematical tools they will need in their further Business education. In addition to his employment at The Hague University, Wouter is currently expanding himself as a photographer. In his free time he also likes to hike and cycle long distances. In 2012 he hiked the last 1000 kilometres of the camino to Santiago de Compostella.